Guide to Improving Pipeline Safety by Corrosion Management

Natural Gas and Hazardous Liquid Transmission Pipelines

NACE
INTERNATIONAL

1440 South Creek Drive
Houston, Texas 77084

NACE INTERNATIONAL

THE CORROSION SOCIETY

Library of Congress Cataloguing in Publication Data

Hevle, Drew.
Guide to improving pipeline safety by corrosion management / Drew Hevle.—1st ed.
p. cm.
Includes bibliographical references and index.
ISBN 1-57590-243-5
1. Petroleum pipelines—Maintenance and repair. 2. Petroleum pipelines—Corrosion.
I. NACE International. II. Title.
TN879.58.H48 2011
665.7′440289—dc23 2011040780

ISBN: 1-57590-243-5

NACE PRESS

NACE International
1440 South Creek Drive
Houston, Texas 77084
http://www.nace.org

Contents

Foreword

Pipelines are the safest form of transportation, and a critical part of our country's infrastructure. Pipelines play a significant role in the delivery of energy to everyone reading this and likely will for the span of each reader's entire lifetime. Pipelines are playing an increasing role in nontraditional areas, including alternative fuels such as biofuels and environmental activities such as carbon sequestration. Pipelines are here to stay and are safe. That being said, there is certainly room for improvement in pipeline safety. Ultimately society dictates the level of public safety expected. Society balances the pressures of higher safety standards with the additional costs that those standards typically entail and lets industry know indirectly through their government representatives, the media, and increasingly, directly through social media, litigation, and activism. Often the government responds to society's demand for higher safety standards by issuing new regulations for minimum standards. Sometimes the drivers for increases in levels of safety are technologies becoming more affordable. In 1984, New York State passed the first U.S. law requiring seat belt use in passenger cars. Seat belts today save over 11,000 lives every year in the U.S. alone. Seat belts evolved into passive restraints including airbags, anti-lock brakes, stability control systems, and many other safety systems. Today most of us can't even imagine a time when driving without securing our children in child safety seats was common practice.

Pipeline safety in the United States has evolved in a form of punctuated equilibrium. The stasis has been broken up by rare and rapid events driven (for the most part) by pipeline failure incidents. Congress created the Office of Pipeline Safety (OPS) in 1968 to oversee and implement pipeline safety regulations. The first statute regulating pipeline safety was the Natural Gas Pipeline Safety Act of 1968, and Congress later added liquid pipelines to the statute in the Pipeline Safety Act of 1979. These two events were largely driven by a natural gas explosion in 1965 in Natchitoches, Louisiana, that killed 17 people. The cause of the failure was stress corrosion cracking, a failure mechanism that had not been attributed to pipeline failures before this incident. NACE International played a part by developing and publishing its first standard, SP0169 (known as RP0169 at that time), "Control of External Corrosion on Underground or Submerged Metallic Piping Systems," that the corrosion control portions of the new Pipeline Safety Regulations were in part based on. More recent pipeline failures such as those in Bellingham, Washington, and Carlsbad, New Mexico, drove the Pipeline Safety Improvement Act of 2002, and recent failures in the Gulf of Mexico, in Marshall, Michigan, and in San Bruno, California, will certainly mandate more changes to pipeline safety regulations and standards.

Why do pipeline failures generate such notoriety? Motor vehicle accidents kill more than 45,000 people per year in the U.S., but a fatal car crash isn't national news like a fatality caused by a pipeline failure. Incidents related to power lines are much less notorious, and typically around 100 people in the U.S. are killed every year, whereas many fewer people are killed in pipeline incidents. One reason is the record of success that pipeline safety has had to date, similar to the affliction of the airline industry, where the expectation of safety is so high that any incident at all becomes newsworthy. Another reason may be that power lines are a hazard that can be seen and understood, but many people are not aware of the buried pipelines near them.

Much of the low-hanging fruit of pipeline safety has already been plucked. There is always room for improvement, but there are no easy answers, and no paradigm-shifting technologies on the immediate horizon. Improvements in pipeline safety will have to be made by small improvements in a large number of areas. These types of improvements lend themselves to comprehensive programs, to ensure that optimum decisions are made at each stage of the pipeline's life. Pipeline Corrosion Management as presented in this document is intended to provide a comprehensive process for ensuring that each decision related to managing corrosion on pipelines is fully considered. Pipelines are playing an increasing role in our energy infrastructure, society's expectations for public safety are ever increasing, and continuous improvement in managing threats to pipeline integrity such as corrosion is essential.

Drew Hevle

Manager, Corrosion Control
El Paso Corporation

TG 370 Chair

Preface

Considering the investment that owners make to build, maintain, and operate carbon steel onshore transmission pipelines, hereinafter referred to as "pipelines," as well as the risk and hazards of a pipeline failure, it is imperative that owners manage threats to pipeline integrity. A principal threat to pipeline integrity is corrosion. Transmission pipelines are primarily constructed of carbon steel, and corrosion is a time-dependent threat that causes the integrity of transmission pipelines to deteriorate if left unchecked. To maintain pipeline integrity, a concerted design, maintenance, and repair program addressing corrosion threats should be implemented.

The primary goal of this guide is to provide pipeline industry managers and nontechnical personnel with an understanding of the basic issues and requirements for managing corrosion threats to onshore transmission pipelines. It is understood that most owners have integrity management programs in place that deal at some level with corrosion threats, and this guide is not intended to replace these programs but to supplement those programs by providing a foundational understanding of the following concepts:

- Basic corrosion threats to transmission pipelines,
- Risks posed by corrosion threats,
- Corrosion mitigation strategies,
- Assessment methodologies,
- Identifying and implementing remediation activities, and
- Methods for evaluating the effectiveness of a program.

This guide is not intended to provide step-by-step instructions for managing pipeline corrosion; instead it outlines and identifies the essential components of a Pipeline Corrosion Management Program and presents considerations and decision-making tools for developing and implementing a program. Owners can develop their own programs based on their needs and can leverage this guide, and corresponding references, as available resources.

This guide has been prepared by Task Group 370, “Pipeline Corrosion Management,” which is administered by Specific Technology Group (STG) 35, “Pipelines, Tanks, and Underground Systems.” This guide is published by NACE under the auspices of STG 35.

SECTION 1

Introduction

With origins dating back to approximately 1,000 B.C. (with the Phoenicians first using a form of pipeline to transport water), pipelines have become one of the most reliable, cost-effective, and safest modes of transporting vital commodities. Today, an elaborate network of pipelines can be found in nearly every developed country around the world, with more than 1.35 million miles (2.17 million km) of oil and gas pipelines worldwide. This vast network helps deliver more than 50% of the world's energy supply.[1] Ensuring the integrity of this infrastructure is paramount to delivering this critical energy source to its various users and the communities that depend on them.

Pipelines are recognized as being by far the safest mode of transporting natural gas and hazardous liquids, with fatalities resulting from accidents on pipelines occurring much less often than accidents caused by transportation of goods by truck or barge. Fires or explosions are 35 times more likely per barrel of oil transported by vehicle than when oil is transported by pipeline.[2] However, despite being one of the safest modes of transportation, pipelines are subject to a variety of threats and accidents that can result in property damage, injury, loss of life, or environmental damage. Corrosion remains a constant threat to the integrity of pipelines, accounting for over 20% of all significant transmission incidents in the U.S.[3]

Corrosion attacks the steel walls of a pipeline, reducing its capacity to retain pressure. If corrosion is left unchecked, the resulting failure of the pipe wall can rapidly release oil or gas into the environment, causing fire and environmental damage. Figure 1 shows a pipeline failure caused by corrosion.

This guide presents a defined process for the management of corrosion specific to pipelines carrying natural gas and hazardous liquids. To help define

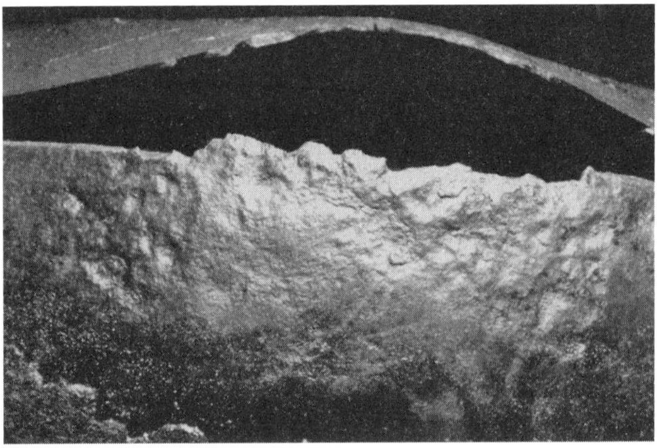

FIGURE 1 Picture of a Pipeline Failure Caused by Corrosion

this corrosion management process and provide context for the process, this guide provides a basic overview of the following topics:

(1) A high-level overview of the corrosion threats to onshore gas and hazardous liquid transmission pipelines;
(2) Methodology and process workflow along with economic considerations to assist in the corrosion management decision-making process;
(3) Approaches to evaluate the risks posed by corrosion threats; and
(4) Key mitigation, assessment, and remediation strategies to preserve pipeline integrity and reduce the risk from corrosion threats.

SECTION 2

An Overview of Corrosion Concerns on Pipelines

Pipelines are the safest and most economical way to transport valuable gas and liquid commodities such as natural gas, crude oil, refined products, and liquefied natural gas. These commodities may be hazardous when unintentionally released into the environment, with a potential for devastating impact to the public, to the operator's employees, to wildlife, property, and to the operator's reputation. Taking steps to prevent such releases and improve reliability is both economical and, in most countries, mandated.

Specifically, the consequences of a corrosion failure on a transmission pipeline include:

- Injuries and loss of life and the financial, political, and personal strain caused by the tragedies. The U.S. gas transmission and hazardous liquids pipeline industry reported a total of 10 serious incidents (those involving death or hospitalization) caused by pipeline leaks or failures from 1990 through 2009.
- For liquid pipelines—environmental damage accompanied by significant environmental fines, cleanup costs, and public outrage from the release of hazardous substances into potable water sources and sensitive environmental areas.
- Liability for damage to real property. Reported incidents in the U.S. from 2005 to 2010 totaled $2.3 billion in property damage.
- Damage to public relations. Large incidents are well publicized in the media and often inspire regulatory and/or legislative response from government bodies. This has economic consequences for the entire industry in addition to the operator responsible for the incident.
- Loss of product and potential loss of revenue that can result from the disruption of service.

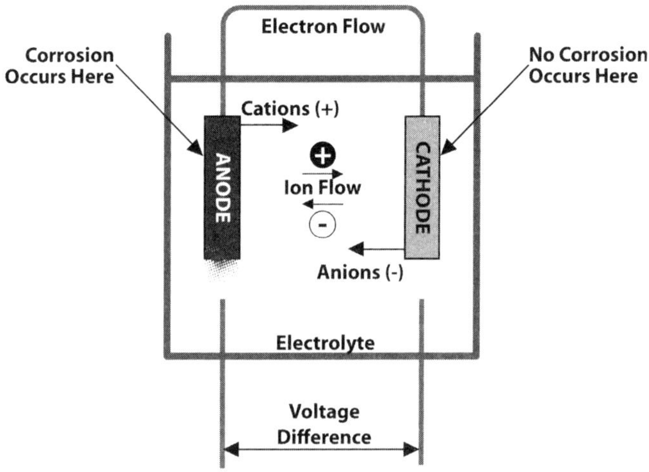

FIGURE 2 Diagram of Corrosion Cell[5]

Because of the unacceptable nature of these consequences, a significant investment has been made in the mitigation, prevention, and control of corrosion in an attempt to avoid costs associated with pipeline failures. For example, the cost of corrosion-related failures on U.S. pipelines constitutes less than 10% of expenses, while capital and operating expenditures for these preventive and control measures is more than 90% of the total expenses, representing approximately $4.9 to $7.7 billion annually.[4]

A Basic Overview of Corrosion

Corrosion is the deterioration of a material, usually a metal, resulting from a chemical or electrochemical reaction with its environment. Most forms of corrosion occur through the formation of an electrochemical cell consisting of the following four elements:

(1) Anode
(2) Cathode
(3) Electrolyte
(4) Metallic (electronic) path

Figure 2 diagrams the corrosion process. Metallic ions (such as Fe^{2+}) leave the metal at the anode and enter the electrolyte. Ions carry the positive charge through the electrolyte to the cathode, while the electrons created at the anode travel through the metallic path to the cathode. The reaction at the cathode completes the circuit. Metal loss of the material occurs at the anode, as a result of ions leaving the metal.

The anode and cathode are defined by a difference in electrical potential. This can be caused by local differences in the electrolyte, microscopic variations in metal structure, different types of metals being joined together, an external source of imposed current (rectifier), or other sources of electrical energy. This diversity leads to a wide variety of corrosion phenomena. Common corrosion mechanisms relevant to transmission pipelines are covered below; for a brief description of other types of corrosion, visit http://events.nace.org/library/articles/corrosion101.asp.[5] Corrosion on pipelines is generally categorized as external corrosion, internal corrosion, or environmentally assisted cracking (EAC).

External Corrosion

Even relatively dry soils may contain enough moisture to serve as the electrolyte in a corrosion cell. External corrosion of buried pipelines is influenced by a number of factors such as external coating condition, ion concentrations in the soil, soil particle size, soil moisture content, proximity to other buried structures, stray direct current (DC), and even induced alternating current (AC) from high-voltage electric lines. External corrosion may be general corrosion, or may take the form of *pitting*, a phenomenon in which localized areas of corrosion called "pits" are surrounded by large cathodic areas of little to no corrosion. The main methods of controlling external corrosion are barrier coatings and cathodic protection (CP; both covered in detail in Section 6).

Internal Corrosion

Most materials transported by pipelines are nonconductive, water-free, and cannot act as the electrolyte portion of the electrochemical cell. However, impurities can be entrained in the fluid stream, leaving small amounts of water and bacteria that can accumulate in the pipeline and enable corrosion.

In dry gas pipelines, an electrolyte is not normally present. Water is present as an impurity in gas streams as they are produced from the well but is removed chemically or physically before transportation in a pipeline. However, water may subsequently be introduced to the system through periodic flooding upsets, maintenance activities, hydrostatic pressure testing, hereinafter referred

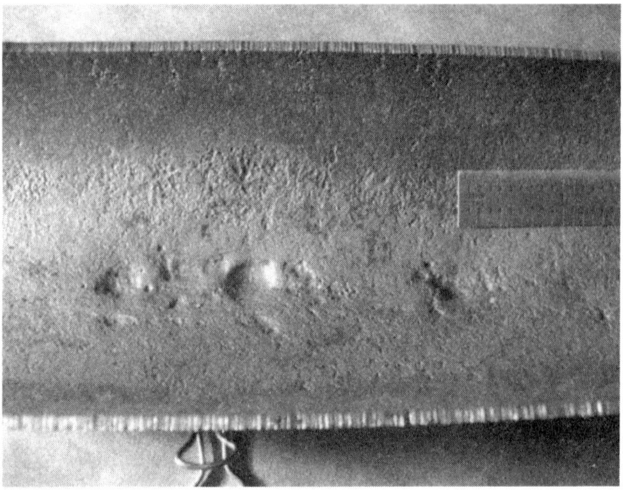

FIGURE 3 Picture of Internal Corrosion on a Pipeline

to as "hydrotesting," or condensation. The water can accumulate and persist at low points and allow internal corrosion (see Figure 3). The presence of corrosive constituents (such as the acid gases CO_2 or H_2S) in the product stream may exacerbate the problem by making the water more corrosive.

In liquid pipelines, the transported fluid may itself be corrosive or water may exist in a two-phase mixture with the fluid.

Environmentally Induced Cracking

Environmentally induced cracking (EIC) occurs when chemical action of the environment (internal or external to a pipe) and loading assists the growth of cracks. This can occur if the crack tip becomes the anode of a corrosion cell, with crack tip dissolution lengthening the crack. Unlike pitting corrosion, which can occur in a wide variety of environmental conditions, EIC mechanisms typically occur only under specific circumstances.

There are several types of EIC. Types relevant to pipeline transmission include the following:

Stress corrosion cracking (SCC) occurs when a susceptible steel material is exposed to tensile stress, the absence of CP, and specific environmental factors. Carbon steel experiences SCC in the high-pH regime when exposed to a solution of carbonate and bicarbonate ions with a pH between 8 and 11. It can also experience near-neutral SCC when exposed to groundwater containing CO_2 at a pH of 5.5 to 7.5. High-pH SCC development is also sensitive to temperature, experiencing higher crack growth rates at temperatures above 100 °F (38 °C).[6,7]

Further information on SCC can be found in NACE Publication 35103.[8]

Hydrogen-induced cracking (HIC) occurs when chemicals that prevent the combination of hydrogen atoms into molecules are present. The cathode of a corrosion cell produces atomic hydrogen as part of the chemical reaction. If it does not immediately form gas molecules, this hydrogen diffuses through the steel wall. Hydrogen atoms within the steel have a number of harmful effects, including embrittlement and accumulation of gas blisters, especially in laminations. The most common chemical that accelerates this process is H_2S, and as a result HIC is sometimes also called stepwise cracking.[9,10]

Further information on HIC can be found in ANSI[(1)]/NACE MR0175/ISO[(2)] 15156-2.[11]

Corrosion fatigue (CF) results from a combination of a cyclical stress and a corrosive environment. Corrosion speeds up the processes of both fatigue crack initiation and propagation. In addition, fatigue can act as a crack initiation mechanism for SCC in pits.

Other Corrosion Mechanisms

Electrical energy sources near a buried pipeline can act as drivers for a corrosion cell. For instance, overhead power lines or electrified rail lines are examples of common energy sources that can cause corrosion from induced stray AC or DC, often called interference corrosion or stray current corrosion. Other buried structures with CP systems can also cause stray current corrosion in a similar manner. Corrosion usually occurs where these unwanted currents exit the pipeline.

Galvanic corrosion occurs when different metals are electrically connected to each other in a common electrolyte. For instance, carbon steel pipe is often connected to a copper grounding grid in a plant environment. The pipeline acts as an anode protecting the copper grounding system, the cathode. Another

---

(1) American National Standards Institute (ANSI), 25 West 43rd St., 4th Floor, New York, NY 10036.

(2) International Organization for Standardization (ISO), 1 ch. de la Voie-Creuse, Case postale 56, CH-1211 Geneva 20, Switzerland.

common example in which galvanic corrosion may occur is when a new pipe is welded to an existing pipeline with the incorrect welding consumable, especially if the weld is anodic with respect to the pipe.

Selective-seam corrosion has been known to occur on pre-1970 flash-welded and DC electric resistance–welded (ERW) pipe. During welding, the lack of consistent heating allows the oxide impurities to remain in the weld centerline, which may result in accelerated corrosion of the longitudinal weld seam.

Microbiologically influenced corrosion (MIC) occurs when biological activity contributes to corrosion on a pipeline. MIC can occur on pipeline internal or external surfaces when water is present, and can be identified by the presence of certain colonies of bacteria. The main types of bacteria that contribute to corrosion are colonies of acid-producing bacteria, sulfate-reducing bacteria (SRB), and slime-forming bacteria. Organic acids and sulfides speed up the chemical reactions associated with corrosion, while the slime formers create concentration gradients on the surface of metal.

SECTION 3

Pipeline Corrosion Management

The Pipeline Corrosion Management process provides a common methodology for reducing the risks caused by corrosion. Through each iterative application, the process needs to continually improve the operator's prevention and mitigation safeguards against corrosion failure. In addition, the process should provide an effective and economical approach to corrosion mitigation, assessment, and remediation. Many of the elements described below are also embodied in existing regulations, code requirements, and industry guidelines.

Pipeline Corrosion Management Process

The flow chart in Figure 4 outlines a process for pipeline operators to manage the risk from corrosion threats. Pipeline corrosion management is a continuous improvement process requiring periodic review and update to ensure programs are effective, threats have been properly identified, risks have been properly evaluated, and new tools, technologies, and current best practices are being implemented. A brief summary of the process is provided below:

(1) The operator first identifies the corrosion threats that are applicable to the pipeline. The process continues for corrosion mechanisms considered to be a threat, continually monitors these threats, and periodically reevaluates all threats to see whether new corrosion threats are applicable to the pipeline.
(2) The risk posed by these threats is then evaluated.
(3) To protect public safety and preserve key assets while considering the risks, costs, and benefits of a particular course of action, the pipeline operator should determine whether mitigation is needed, the available

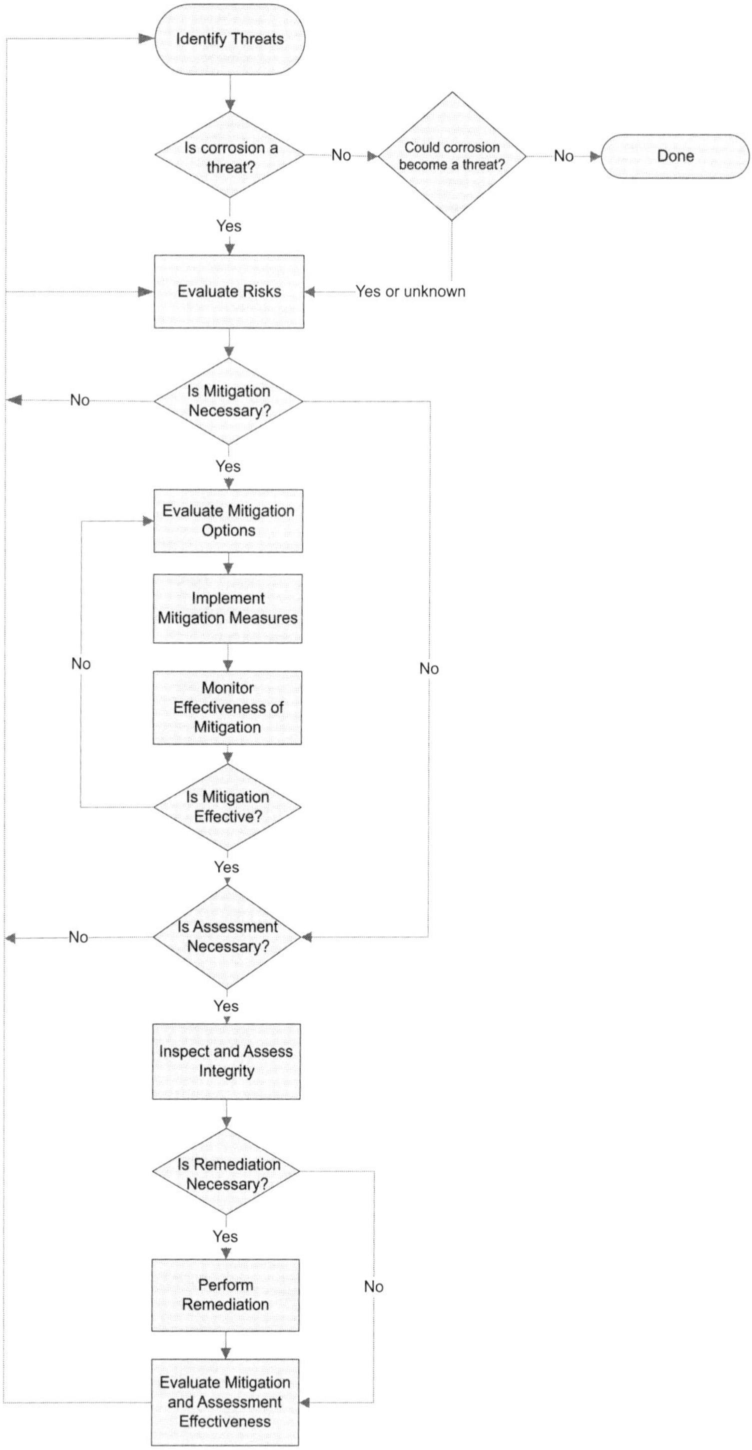

FIGURE 4 Corrosion Management Process Workflow

options for mitigation, and implement the necessary mitigation activities to reduce risk to an acceptable level.

(4) A monitoring program is then implemented to track the effectiveness of applicable mitigation activity (if any). If mitigation is deemed ineffective, options should be reevaluated for improved alternatives.

(5) The pipeline operator performs the necessary assessments to determine the presence and extent of a threat from corrosion damage over the history of operations of the pipeline. Some minimal specification for the frequency of assessments may exist in regulatory requirements, but should be more rigorous and further specified in the operator's integrity management plan.

(6) If an assessment discovers corrosion, further mitigation or remediation should be performed as described in the operator's operation and maintenance (O&M) plan.

(7) A program effectiveness evaluation must be performed periodically. This involves integrating information about program effectiveness, new technologies, and emerging threats to improve the corrosion management program.

(8) Threat identification/monitoring should be performed periodically. New threats and results of mitigation monitoring, previous assessment, and remediation activities should be incorporated into the risk management program.

SECTION 4

Corrosion Threat Assessment

Identification of corrosion threats to a particular pipeline requires an evaluation of the design, operational, and environmental factors. Subject matter experts (SMEs) may be used to help supplement documented information, especially when documentation is incomplete. Using guidance provided by the operator's historical performance, the U.S. Department of Transportation's (DOT) Pipeline and Hazardous Materials Safety Administration (PHMSA)[(3)] and other regulatory authorities, and other accident/leak reports on pipelines with similar design and operating characteristics, regulatory requirements, and industry standard references (such as those from NACE, ASME,[(4)] and API[(5)]) combined with sound engineering judgment, a decision can be made for the inclusion or exclusion of certain corrosion threats.

For buried and aboveground piping, external corrosion is usually assumed to be a threat for all situations.

Internal corrosion is similarly identified and widely accepted as a threat to pipelines. Water content must be monitored and mitigated unless careful specifications and monitoring of the pipeline are in place.

SCC is assessed for risk as a possible threat if the following conditions are met:[12]

(1) The operating stress exceeds 60% of the specified minimum yield strength (SMYS) of the pipeline at some time in its life;
(2) The age of the pipeline is greater than or equal to 10 years; and
(3) The coating type is other than fusion-bonded epoxy (FBE).

---

(3) U.S. Department of Transportation (DOT) Pipeline and Hazardous Materials Safety Administration (PHMSA), 1200 New Jersey Ave. SE, Washington, DC 20590.
(4) ASME International (ASME), Three Park Ave., New York, NY 10016-5990.
(5) American Petroleum Institute (API), 1220 L St. NW, Washington, DC 20005-4070.

TABLE 1 Sources for Threat Identification Considerations

Code/Standard	Applicability
U.S. CFR 192.917[13]	How an operator identifies threats on gas transmission pipelines in the U.S.
U.S. CFR 195 Appendix C[14]	Guidance for implementation of an integrity management program for hazardous liquid pipelines
ASME B31.8S[10] Section 2.2, Tables 6, 9	Managing system integrity of gas pipelines; identifies key applicable threats to transmission pipelines
API 1160 Section 5.2[15]	Managing system integrity for hazardous liquid pipelines—framework elements

All other environmentally assisted cracking (EAC) threats should be reviewed, evaluated, and identified by integrity management and corrosion personnel.

Table 1 provides key sources for threat identification and transmission pipeline integrity.

SECTION 5

Economic Considerations and Risk Evaluation

Economic Considerations

Most jurisdictions require a certain level of pipeline corrosion management by law. These minimum standards are designed to protect public safety. Most operators find it prudent to go beyond the minimum requirements. These operators are faced with a difficult question: how best to allocate and prioritize resources and money in the prevention and detection of corrosion? The tools covered in this section can help to provide answers to such questions.

As part of the corrosion management process, operators should perform an analysis of the tradeoffs between corrosion risks, upfront capital or operation and maintenance (O&M) investments, and deferred or recurring expenses. This balance may be analyzed using several common economic and decision tools. Historical NACE Publication 3C194[16] on Economics of Corrosion (see Appendix B) discusses several of these tools, including

- Internal Rate of Return (IROR)
- Discounted Payback (DPB)
- Net Present Value (NPV)
- Benefit Cost Ratios (BCR)

Each of these methods relates future expenses to present expenses.

The value associated with corrosion prevention and risk reduction strategies, including mitigation, assessment, and remediation activities, can best be evaluated using these economic decision tools. The process can range from a simple cost analysis to a more elaborate, complex analysis using spreadsheets or sophisticated software programs detailing the costs involved, resultant values, assumed interest rates, and expected value for each strategy. An NPV analysis (see Appendix A) is often viewed as the optimal method for evaluating a strategy, particularly when a series of actions are expected to incur costs and/or cash flows over a number of years in the future. Sometimes using a qualitative or a simple three-year payback method may be sufficient.

It is important that operators ensure no costs are overlooked in other budget areas. Costs may include both O&M expenses and/or capital outlays for a set of mitigation, assessment, and remediation activities to drive a certain level of reduction in corrosion risk. The benefit in terms of risk reduction may be evaluated and compared with the costs expected to be incurred for an economically justified decision. Appendix A also provides a simple overview and example of a cost-benefit analysis using NPV. More information on techniques for evaluating the costs and technical benefits in evaluating economic decisions of corrosion control is provided in NACE Publication 3C194 (Appendix B).

Risk Evaluation

Evaluating risk is an extremely useful tool and key step in a pipeline corrosion management process. The evaluation can be used to help determine the value in implementing corrosion prevention strategies. In addition, a risk evaluation can help to prioritize corrosion management activities. Resources can be targeted at the highest risk segments and those representing the largest potential losses to an organization. A variety of methods and tools exist for evaluating the risk of corrosion threats. These methods range from very complex, quantitative, and analytical-based models using sophisticated software tools to simpler models based on subject matter expert (SME) scoring.

Risk is composed of the following elements:

(1) The likelihood of an event.
(2) The consequence if that event were to occur.

Likelihood can be expressed as an actual probability between zero and one, or as a "relative score" with numbers assigned to descriptors such as "very likely" (high score) or "very unlikely" (low score). Likewise, consequence can be expressed in numerical form based on predicted costs (e.g., property damage, lost revenue, and/or repair/cleanup costs), or as a qualitative result such as "minor" to "severe."

Regardless of the methodology, the first step is to gather design, operational, and physical data. This data collection effort may be a simple process of reviewing readily available data in geographic information system (GIS) alignment sheets, or consist of an intensive effort collecting data that may be spread throughout disparate systems and facilities—or may be undocumented or in paper records (see NACE SP0507[17]).

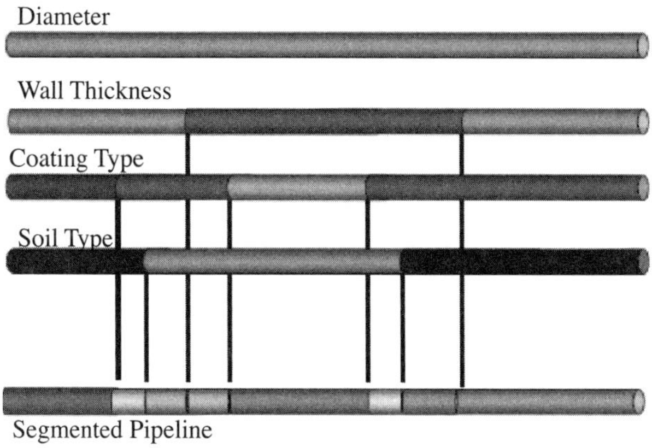

FIGURE 5 Example of Pipeline Segmentation

Because pipeline risk factors are likely to change over different sections or regions of the pipeline, segmenting the pipeline based on similar attributes and those factors likely to contribute to a specific corrosion threat could lead to a more useful risk evaluation and more focused mitigation and assessment activities. Figure 5 provides an example of dynamic segmentation that looks at a select grouping of attributes that are likely to impact risk along the pipeline and how those factors can be segmented for a risk analysis.[20]

The NACE direct assessment (DA) standards,[17,19,20,21,22,23] API 1160,[15] and ASME B31.8S provide detailed lists of minimum data elements that may impact the likelihood and/or consequence factors of pipeline integrity threats. Although not exhaustive, Table 2 provides a brief overview of some of the factors that influence the likelihood or consequence of components of risk. More detailed lists with further information on each item to consider collecting are provided in NACE SP0502,[20] NACE SP0204,[12] NACE SP0206,[19] NACE SP0507,[17] API 1160,[15] and ASME B31.8S.[11]

The different approaches to evaluating risk to a pipeline can be grouped into four basic categories:

- SME
- Scenario-Based
- Relative Assessment
- Probabilistic

Table 3 provides a basic overview of each category with further detail found in ASME B31.8S and *Pipeline Risk Management Manual*.[24]

TABLE 2 Consideration Factors for Evaluating Risk

Key Factors Contributing to Likelihood of Failure	Design, Material, and Construction Data • Pipe route/coordinates • Pipe grade • Pipe diameter • Pipe wall thickness • Pipe (seam) type • Maximum operating pressure (MOP) and safety factors • Manufacturer and production date • Construction date • Joining/bending method • Weld quality and inspection • Coating type and condition • Cathodic protection type and condition • Soil/backfill type • Appurtenances • Weight hold-downs • River/lake/water crossings • Shared right of way/utility crossings • Corrosion rate O & M data • Gas composition • Supervisory control and data acquisition (SCADA) and leak detection • Operating temperature • Flow rate • Product quality • Ambient temperature • Operating pressures, cycles, and profile • Supervision of excavations • Condition of right of way • Pipe-to-soil, close-interval survey (CIS), and CP data • Repairs • In-line inspection (ILI) and other assessment/inspection results • Hydrotest data • Corrosion monitoring and coupon data
Key Factors Contributing to Consequence Factors	• Pressure/release factors • Product type/composition • Valve locations • Proximity to populated areas • Potential for endangering human safety • Hazardous liquid lines: Proximity to drinking water, recreation water areas, proximity to other water use/waterways, farms, parks and forests, commercial fishing waters, sensitive areas, and other important areas

	• Potential for releases in canyons • Potential for environmental impact (air, soil, and water) • Potential for fire • Potential for financial losses • Human safety, injury, and fatality • Damage to air, soil, and water • Functional losses, legal expenses, fines, and punitive damages • Costs of alternative drinking water supply, down time, and fuel shortages

TABLE 3 Risk Assessment Approaches

(SME)	• An SME-based approach leverages resident knowledge of key personnel and expertise within the organization to evaluate the risk of corrosion threats. • A risk scoring methodology that assigns a likelihood score for a failure and determines the consequences based on information collected from the SMEs for each pipeline segment is typically devised. • SMEs are sometimes the best source of information on a pipeline. • Individuals can be biased, and different SMEs may give different assessments of the same equipment.
Scenario-Based	• A scenario-based approach involves investigating and tracing backward the series of events leading to a failure or fault event and evaluates the level of risk associated with each event. Event, fault, and decision trees are typically created to assist in this methodology.
Relative Assessment	• A relative risk ranking approach builds on pipeline operation and maintenance experience, incorporates more extensive and quantitative data representative of corrosion threats, and defines an algorithm to segment the pipeline and create a relative risk score based on these data. The resultant score for each segment identifies risk on a relative basis to compare pipeline segments, which can then be used to better target mitigation, assessment, and remediation activities. • Although more complex and intensive than an SME-based approach, this approach is typically less subjective because it is more reliant on quantitative data.
Probabilistic	• A probabilistic risk assessment is one of the most complicated risk evaluation methodologies because it is based on performing a rigorous statistical and mathematical analysis to define a probability of a particular corrosion failure. A statistical analysis is performed on each event leading to a failure with the total probability of the failure based on the combined probability of the events occurring.

The advantages and disadvantages of each approach should be weighed to determine the optimal approach and alignment with the best fit for an operator's integrity management program. Key factors include data availability as well as time and costs to gather the required information and implement the risk evaluation process. Risk evaluation approaches may be combined or mixed provided there is a sound rationale. Essentially these approaches are methods to prioritize information for decisions and should not automatically determine the replacement, mitigation, and prevention activities.

SECTION 6

Mitigating Corrosion Risk

Integrity risk can be reduced by limiting either of its components: likelihood or consequences of failure. Consequences are generally limited by physics and encroachment, over which the operator has limited control. Therefore, operators typically focus on reducing the likelihood of failure. There are several methods that can be used throughout a pipeline system to reduce the likelihood that corrosion will occur. Which methods are employed depends on the expected type of corrosion, priority, cost, availability of choices, and sometimes regulatory requirements.

Operators may choose mitigation methods based on the cost to achieve an expected level of reduction in failure likelihood. The value associated with a particular reduction in failure likelihood can be evaluated using the economic decision tools described in Section 5. Other inputs to the decision process may come from regulatory constraints, company standards, industry consensus standards, and known industry best practices. Scenario planning can also be incorporated into the analysis to compare strategies and determine the optimal combination of actions to address each specific corrosion threat.

The following sections provide an overview of some common mitigation strategies, guidance on selecting the appropriate strategy, and some additional resources for further information on each strategy.

External Corrosion Mitigation

NACE SP0169 (formerly RP0169)[25] defines several potential mitigation activities, primarily in the design phase, to help in the prevention of external corrosion. At the design phase, external corrosion mitigation activity can be divided into four primary categories:

(1) Materials selection and design
(2) Protective coatings (painting and buried coatings)

(3) CP
(4) Modification of environment (control of backfill)

Various options may be evaluated for these mitigation factors based on improvements to the operating capacity, changes in expected life of the pipeline, and reduction in risk relative to the capital and operating expenditures to implement these measures. Through a combination of mitigation techniques, an optimal point of safe operation, costs, and life expectancy of the asset can be attained.

For example, it may be possible to apply CP without a pipeline coating and obtain enough current density on the pipeline to provide sufficient corrosion protection, but this will likely lead to significantly increased costs in the form of rectifiers, anode beds, and increased power consumption costs, as well as a potential for creating stray current issues for neighboring or lateral facilities. However, if a coating were applied along with CP, current demand from the CP system would be much lower.

As described in detail in NACE SP0169, the following considerations should be carefully analyzed with regard to impact on external corrosion during design:

- Material type/grade: The use of corrosion-resistant materials reduces the rate of corrosion but may increase cost. In addition, pipelines are normally designed with thicker walls than are necessary based on the design factor. This allows for a reasonable response time to address deficiencies or remediate anomalies discovered.
- Points for electrical isolation and isolation devices: casings, facilities, and other electrically influencing areas should be carefully planned to avoid an unwanted impact or shielding of the CP.
- Connecting and positioning of CP and corrosion control test stations: The type, quantity, and placement of anode beds or other CP systems in different soil systems have a direct impact on the amount of protection to the pipeline. The accessibility of test stations impacts the ability to monitor and ensure that a sufficient level of protection is provided. Test stations should be provided at enough locations and installed correctly to better facilitate CP testing and monitoring (see NACE SP0169 for more detail).
- Coating selection (see section on *Protective Coatings*): Coatings are the first line of defense against corrosion because coatings help prevent contact with an electrolyte, preventing completion of a corrosion cell.

- Construction practices: Leveraging industry best practices can help verify preventive systems are properly installed and help avoid damage to coating and electrical circuit systems during construction.
- Backfill material and application: Careful selection is necessary to verify that the properties are adequate for the environment (soil stress, backfill material) and operational characteristics (temperature).

Materials Selection and Design

External corrosion control should be a primary consideration during the design phase of a piping system. Design factors that minimize external corrosion and provide tolerances for a certain amount of corrosion while maintaining safety factors should be evaluated against the respective construction costs of the pipeline. In addition, pipelines typically are subject to stringent requirements in codes and regulations for compliance at all times with these safety factors and hydrotest performance criteria. Other benefits of stringent design factors include lower stresses in the pipe wall, and increased resistance to other threats such as third-party damage and outside forces. Certain design factors (such as installing CP test points or casings) can also affect the ability to cost-effectively monitor external corrosion on the pipeline and should be considered.

Protective Coatings

External coatings, hereinafter referred to as “coatings,” are the first line of defense against electrochemical corrosion because coatings block exposure to a corrosive environment. Coatings typically contain defects, known as holidays, which occur as a result of application, transportation, storage, and installation procedures. Over time, buried coatings can also break down as a result of interaction with the environment surrounding the pipeline. Concentrated areas of external corrosion can occur at these holidays, and CP is provided to protect those areas where the coating breaks down. Therefore CP and coatings work in synergy to protect buried pipelines.

Coatings must be correctly selected for the expected range of operating conditions. Coatings must also be properly applied and carefully handled during transportation and installation to prevent damage. Various types of coating materials can usually accomplish the desired functions, and an economic analysis can help determine whether a coating is needed and what type of coating is best suited for the application. This analysis should first define the requirements and qualifications of the coating to ensure corrosion issues are properly addressed.

The following parameters are known to have an impact on the performance of a coating and thus should be evaluated and considered in the coating selection:[26]

- Type of environment
- Accessibility of the piping system
- Operating temperature of the piping system
- Ambient temperatures during application, shipping, storage, construction, installation, and pressure testing
- Geographical, soils, and physical locations
- Type of coating on the existing pipe in the system
- Handling and storage
- Pipeline installation methods
- Costs (material, application, durability, and repair)

Buried/Submerged Piping Structures: In addition to protecting the pipeline directly from external corrosion, coatings on buried/submerged pipelines also need to allow for effective distribution of current to holidays. The following characteristics are defined as the principal criteria for coatings on buried or submerged pipelines:[26]

- Effective electrical characteristics
- Effective moisture barrier
- Good adhesion
- Applicability
- Ease of repair
- Resistance to disbondment
- Resistance to chemical degradation
- Ability to maintain constant electrical resistivity over time
- Ability to resist damage during handling, storage, and installation
- Retention of physical characteristics
- Nontoxic to the environment

Table C1 in Appendix C provides a brief description of some common protective coatings as well as some applicable standards and references for these coating types.

For buried or submerged pipelines, the coating type is likely to have an impact on the costs of CP and should be viewed holistically when coating types are evaluated.

Surface Preparation: A properly cleaned and prepared surface prior to coating application is critical for ensuring the durability of the coating. Even the best coatings may fail rapidly if not applied to a surface properly prepared to receive the coating.[26] NACE and the Society for

TABLE 4 NACE and SSPC Surface Preparation Standards

Method of Surface Preparation	Standard Number
White Metal Blast Cleaning	NACE No. 1/SSPC SP 5,[27] ISO 8501-1[28]
Near-White Metal Blast Cleaning	NACE No. 2/SSPC SP 10,[29] ISO 8501-1
Commercial Blast Cleaning	NACE No. 3/SSPC SP 6,[30] ISO 8501-1
Acid Pickling	SSPC SP 8,[31] ISO 8501-1
Brush-Off Blast Cleaning	NACE No. 4/SSPC SP 7,[32] ISO 8501-1
Power Tool Cleaning	SSPC SP 3,[33] ISO 8501-1
Hand Tool Cleaning	SSPC SP 2,[34] ISO 8501-1
Solvent Cleaning	SSPC SP 1[35]
Waterjetting	NACE No. 5/SSPC-SP 12[36]
Industrial Blast Cleaning	NACE No. 8/SSPC-SP 14[37]

Protective Coatings (SSPC)[(6)] have published standards for different levels of surface preparation as shown in Table 4.

Mill-Applied Coatings: There is a need for mill and field coating applications. Coating manufacturers typically specify a thorough application and testing process to optimize the effectiveness of the coating and minimize defects. Generally, better quality control of the process and materials handling as well as better manufacturing productivity may occur in a mill environment; several newer coatings are commonly applied in the factory or mill prior to transport for installation. However, the downside to this approach is the potential for greater ultraviolet (UV) exposure and handling damage during storage and transit to the field. As an example, Figure 6 outlines a typical process for applying a FBE coating in a mill.

Field-Applied Coatings: To get adequate coating coverage of welded joints, select appurtenances, and make repairs on identified holidays, a preparation and application process must also exist for applying coatings in the field. Older application methods for coatings such as coal tar, asphalt, and tape wrap were by machine in the field after joints were welded together and cleaned via wire brush or other method. Some potential problems existed with this approach because of poor bonding from ineffective cleaning and scale interfering with adhesion. More recently, sand blasting and other rigorous cleaning techniques have helped resolve some of these problems.[39] The use of technologies such as field-applied FBE coatings and advanced application techniques such as plural-component spray equipment have also resulted in improved durability.

(6) The Society for Protective Coatings (SSPC), 40 24th St., 6th Floor, Pittsburgh, PA 15222-4656.

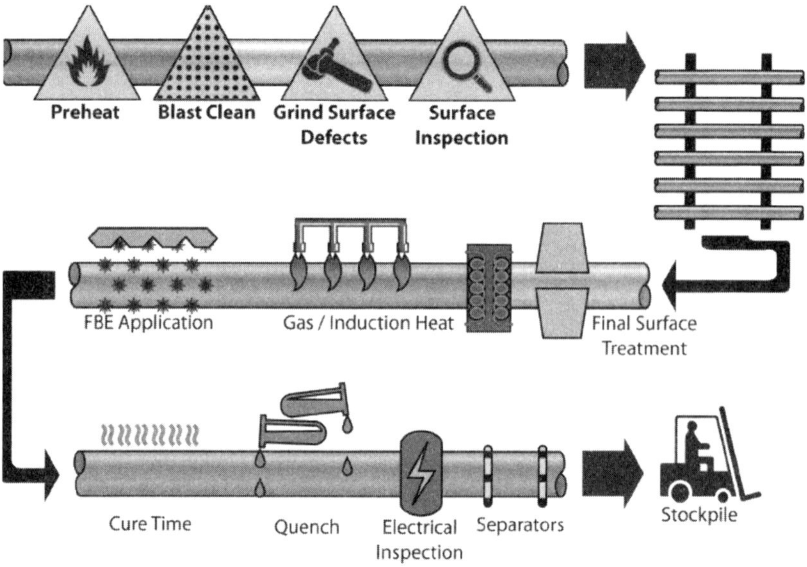

FIGURE 6 Example FBE Mill Application Process[38]

Aboveground Piping: Distinct requirements and considerations are required for sections of exposed aboveground pipeline. Coating is one of the few and typically the most effective defense against external corrosion for aboveground pipelines. Usually the painting or coating of aboveground pipelines may involve multiple layers consisting of a primer, intermediate coat, and topcoat. Special factors should be considered for each coating layer, as shown in Table 5.

Table C2 in Appendix C provides an overview of select coating types, relevant standards, and characteristics of coatings for aboveground piping.

Cathodic Protection

CP is a mitigation method that reduces the corrosion rate of a pipeline by making it the cathode of an electrochemical cell, and when combined with coatings, CP is the most effective and prevalent method for protecting pipelines against external corrosion.[40] There are two types of CP deployed with transmission pipelines:

(1) Galvanic or sacrificial anodes
(2) Impressed current

Galvanic or sacrificial anodes take advantage of purposely establishing a corrosion cell with dissimilar metals that is strong enough to counteract corrosion

TABLE 5 Aboveground Coating Layer, Description, and Considerations

Coating Layer	Considerations/Functions	Example Types
Primer	• Bonding to bare steel surface • Adhesion to intermediate/topcoats • Corrosion suppression (for limited exposure and to retard spread of corrosion to other discontinuities)	• Alkyd or oil • Mixed resin • Resin identical topcoat • Inorganic zinc
Intermediate Coat	• Provide adequate film • Uniform bond between primer and topcoat	• Various
Topcoat	• Initial barrier to the environment • Resistance to chemical, UV, and atmospheric degradation	• Epoxy • Inorganic zinc • Organic zinc • Furan • Urethane • Coal tar epoxy

cells normally occurring on the pipeline. When the pipeline is connected to a more active or sacrificial metal, current discharges through the anode, causing the sacrificial metal to corrode instead of the pipeline.[43]

Impressed current, on the other hand, uses DC from an outside source, such as a rectifier also connected to an anode bed, to provide a potential shift at exposed holidays on the pipeline. When sufficient current causes a voltage potential shift to the exposed pipeline surfaces, the bed typically serves as an anode in a corrosion cell, the typical free corrosion current is reversed at these anodic areas, and corrosion is prevented at these holidays (see Figure 7).[41]

There are advantages and disadvantages with each approach that are briefly summarized in Table 6.

TABLE 6 CP Type Comparison

Galvanic/Sacrificial	Impressed Current
Advantages: • Easily installed • Less costly over the short run for a localized segment/area • More limited power requirements	*Advantages:* • More even current distribution • Ability to cover a greater area/segment of pipeline
Disadvantages: • Extremely localized current distribution • More frequent and higher replacement costs • Difficult to isolate during surveys or indirect inspection	*Disadvantages:* • More expensive installation costs • Power/electricity requirements

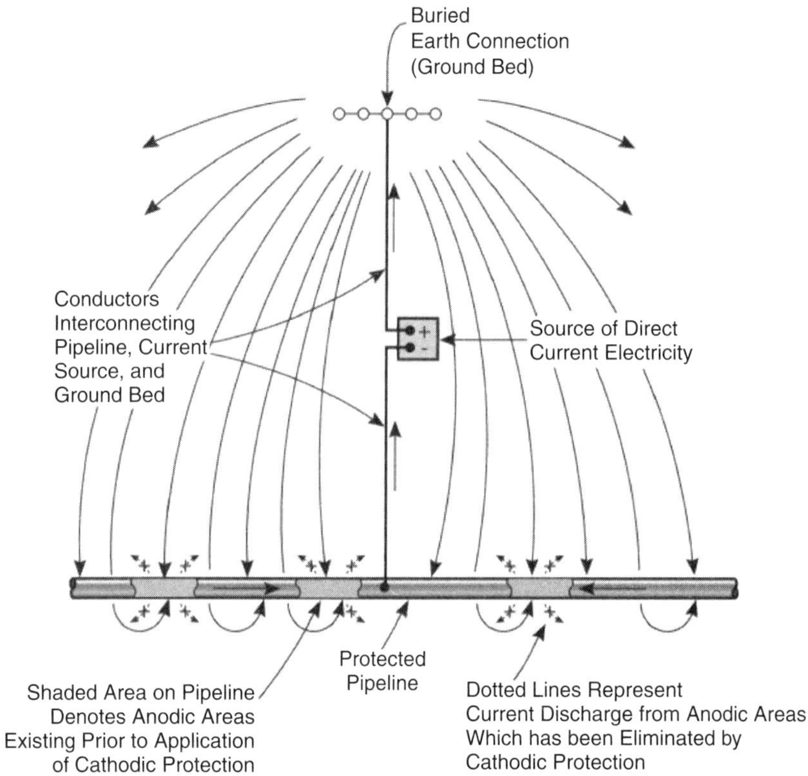

FIGURE 7 Overview of Impressed Current[42]

Ultimately, the current density requirements for the pipeline, or the required amount of current the CP system needs to provide to polarize the pipeline sufficiently to overcome corrosion, combined with total overall costs, should be evaluated to determine the type and details of the CP deployment. Detailed knowledge of the system and selected isolation points are also required to make the determination for the CP design. For example, one industry practice employed in defining CP requirements is to assume 90% to 95% coating efficiency (depending on age and quality of coating type) and ensure 1 to 2 mA of coverage per 0.3 m^2 (3.2 ft^2). Additionally, current requirement testing can assist in determining the levels of current required to provide adequate CP to a segment of piping. In general, galvanic systems should be considered only if the current requirement is less than 100 mA. Several quantitative software tools are also available to assist in the design and selection of CP systems.

The following criteria have an impact on the cost and performance of CP and should be considered when determining the type and/or location of the groundbed:

- Availability of commercial power
- Locations of lowest and highest possible soil resistivity
- Sufficient right-of-way space
- Clearance for installation and construction equipment
- Adequate isolation from foreign metallic structures to minimize electrical interference
- Installation costs
- Depth of groundbed
- Accessibility
- Current requirements
- Soil aggressiveness

Once total system requirements are known based on current demands from each isolated segment, the total costs combined with considerations for the advantages/disadvantages of each approach can be evaluated to make an optimized decision on CP design. Although the initial design and implementation may be capital intensive, recurring operating costs in the form of power consumption and CP monitoring costs as well as equipment replacement costs should be considered.

Because stray currents and other sources of electricity can interfere, negating these mitigation efforts, consideration and strategies should be employed to limit the impact on the pipeline system. These locations may include, but are not necessarily limited to, the following examples:

- Adjacent pipeline systems
- Light rail
- Mining operations
- Electric transmission lines

Appendix A provides a basic example evaluating the tradeoffs and decision-making criteria associated with these sources.

Modification of the Environment (Control of Backfill)

Another mitigation option is to select backfill material that minimizes a corrosive environment during design or excavation. Design engineers should consider corrosion factors as part of the soil-structure interaction and evaluate the benefit to corrosion prevention against the soil procurement, additional design, and deployment costs. Factors that can affect the corrosion rate with regard to soil/backfill include:

- Soil chemical characteristics (non-acidic pH, corrosiveness)
- Ensuring no rocks or rough objects that can cause coating or pipe damage are present
- Proper density to minimize earth movement/provide a rock cushion
- Adequate drainage
- Soil conductivity
- Ease of placement

Ongoing Monitoring of External Corrosion

Monitoring the effectiveness of mitigation activities must be pursued to ensure risks are being reduced as expected and further action is not warranted. Monitoring activity is primarily a function of the type and level of the mitigation activity implemented. Frequency and level of monitoring should consider previous assessment results, risk level, regulatory constraints, and costs associated with monitoring.

Mitigation monitoring tends to have greater O&M costs because it involves the regular collection and analysis of CP performance data. For buried, coated pipelines with CP, monitoring may largely consist of pipe-to-soil potential measurements and indirect surveys to ensure adequate polarization, reducing the threat of active corrosion on the pipeline, and optimizing CP performance. Following is a list of some activities for periodic monitoring of mitigation activities:

- Monitoring rectifier output
- Collecting pipe-to-soil potential measurements
- Periodically installing/monitoring external corrosion coupons
- Performing indirect surveys
- Using a combination of direct assessment methodologies (discussed in detail in Section 7)
- Real-time or periodic monitoring measurements of key performance indicators

Internal Corrosion Mitigation

There are several methodologies for mitigating the risk of internal corrosion control at the design phase that can be grouped into the following categories:[42]

Contractual/Procurement Based: Gas/liquid quality standards and specifications defined in contracts and gathering processes.

Capital-Intensive Measures: Pipeline and system design (e.g., flow conditions, line size changes, deaeration points, liquid removal devices, dehydration, and dew point control) as well as internal linings.

Capital and Recurring/O&M-Based Measures: In the form of chemical additives (corrosion inhibitors, oxygen scavengers, or biocides) and periodic cleaning of the pipeline if contaminants are identified.

Similar to the preventive measures for external corrosion, a combination of these methods is likely to be the most effective approach.

Gas/Liquid Quality Standards and Specifications

One of the most effective methods to minimize the potential for a corrosive environment is to ensure water impurities and corrosive agents never enter the pipeline. Thus, during procurement contract negotiations, consideration should be given to defining standards and specifying contracts with producers and processors to help minimize product impurities that exacerbate internal corrosion.

Typically specifications in procurement contracts limit the amount of impurities present in the source. NACE SP0106[45] defines the following example subset of impurities that are of concern for internal corrosion consideration:

- Bacteria
- Carbon dioxide (CO_2)
- Chloride
- Hydrogen sulfide (H_2S)
- Oxygen
- Solids or precipitate
- Sulfur-bearing compounds
- Water

Added costs of procuring higher-quality product and/or additional processing should be evaluated in context with the risk level, monitoring data, and costs of alternative options.

Pipeline Design Considerations

Similar to external corrosion, internal corrosion control should be a primary consideration during the design, procurement, and construction of a new pipeline. Design factors that minimize corrosion should be evaluated against the respective costs, safety factors, and operating life of the pipeline. Factors to be considered include

- Defining flow velocities and limits such that lower limits avoid suspended impurities dropping out of the gas or liquid and upper limits that avoid erosion-corrosion effects.
- Avoiding intermittent flow so impurities or sediment do not collect in the line.
- Designing line size changes and eliminating dead legs to help eliminate areas where corrosive contaminants can collect.
- Dehydration and dew point control to eliminate water entering the pipeline and eliminating the potential for a corrosion cell.
- On single-phase gas pipelines such as natural gas transmission pipelines, liquid removal devices such as drips, filters, separators, and slug catchers can remove liquids including water. If these devices aren't properly maintained, then the corrosion problem is simply moved from the pipeline to the liquid removal device.
- Process equipment such as acid gas strippers and scrubbers to remove corrosive constituents such as CO_2 and H_2S.
- Deaeration to reduce oxygen content and lower the likelihood of corrosion during the transport of the product.[43]

Internal Linings

Similar to external coatings, an internal lining or coating can help prevent the possibility of a corrosive agent coming into contact with the metallic surface, thus eliminating the potential for the formation of a corrosion cell. Although an internal lining is a valid preventive measure, consideration should be given to potential complexities when the benefits and costs are weighed. These complexities include

- Limited ability to use certain assessment methodologies
- Difficulties in ability to detect coating disbonding and defects once a pipeline is in place
- Hydraulic/flow impact to the pipeline
- Added complexity to cover welds and fabrication
- Additional costs and complexity in application and procurement of piping

Because of the added complexities and costs, internal coating is typically a stronger consideration in aggressive environments such as gathering lines, in-station manifold piping, and other areas where it is uneconomical or less feasible to implement other mitigating actions and is thus not typically applied to transmission lines.[44]

Chemical Additives (Corrosion Inhibitors and Biocides)

Mitigating internal corrosion can also be accomplished by using chemical additives (such as a corrosion inhibitor) that, when added to a hazardous liquid or gas, decrease the corrosion rate of a metal/alloy. Some inhibitors create a passivation layer, or thin film on the surface of the material, that stops access of the corrosive substance to the metal and thus prevents a corrosion cell from forming. Other inhibitors, such as pH buffers and oxygen scavengers, function by chemically altering the environment. A wide array of different inhibitor types are available and can be cathodic, anodic, or both, essentially increasing the anodic or cathodic polarization behavior; reducing the movement or ions to the metallic surface; or increasing the electrical resistance of the metallic surface.[45]

When bacteria are believed to be influencing the corrosion, biocides may be an option to kill the microorganisms to reduce corrosion rates.[45] Before a decision is made on this approach, an understanding of the cause of internal corrosion in combination with a chemical understanding of the pipeline product to ensure proper selection of chemical type, frequency of application, and quantity per application should be factored with evaluation of the respective costs.

Some capital expenditures are incurred in the design of storage vessels, measurement devices, and injection points, but the chemical additives are typically more influenced by recurring O&M expenses for the procurement of chemicals, application of the chemicals, and monitoring of the application effectiveness.

Line Cleaning (Pigging)

Pipeline cleaning devices, commonly referred to as cleaning pigs (see Figure 8), can be useful tools for removal of contaminants within the pipeline and involve a cleaning instrument running through the inside of the pipeline to push contaminants into a receiver or "pig trap."

Economic considerations include the capital- and design-based costs to install valves, bends, launchers, and receivers to enable use of the tool as well as overall operating costs of hiring a crew or contractor to set up and run the tool. Observation of type and quantity of contaminants collected can be useful

FIGURE 8 Illustration of Cleaning Pig and Tools

in monitoring the mitigation effectiveness and can ultimately be used to help determine the frequency of performing cleaning-pig runs.

Internal Corrosion Mitigation Monitoring Activities

Except for the use of corrosion coupons, monitoring and mitigation activities for internal corrosion are primarily O&M-oriented and consist of the following activities:

- Analyzing coupons and probes
- Sampling from traps and chemical analysis
- Visual inspection of the pig
- Periodic nondestructive evaluation (NDE) inspection (magnetic, ultrasonic, radiographic) evaluating wall thickness
- Analyzing cleaning-pig runs (based on frequency and type)
- ILI results

Findings in one area may lead to greater benefit and justification for pursuing additional monitoring activities (e.g., a coupon sample that indicates a high likelihood of corrosion may justify more frequent NDE inspections).

Stress Corrosion Cracking/Environmentally Induced Cracking (EIC) Mitigation

Although limited mitigation strategies exist for preventing stress and environmental corrosion cracking, one principal strategy involves setting design

TABLE 7 SCC Susceptibility Factors

Surface preparation for field coating	Mill scale and wash residual chemicals may create coating disbonding issues that may lead to susceptibility to SCC.
Coating type	High-pH SCC is found under coal tar, asphalt, and tape. Near-neutral-pH SCC is most prevalent under tape wraps but also found under asphalt. Weather conditions during construction also may be important in affecting coating bonding condition.
Coating system and condition	Inadequate coating is a critical factor in SCC susceptibility. The coating system (coating type, condition, etc.) is an important factor not only in determining susceptibility but also the type of SCC that is likely to occur. Because SCC requires coating faults, indications of coating condition might help locate probable areas.
Operating temperature	Elevated temperatures have a strong accelerating effect on high-pH SCC. For near-neutral-pH SCC, temperature also has an accelerating effect on crack growth rate, but elevated temperatures beyond 160 °F (71 °C) can contribute to coating deterioration.
Stress levels and fluctuations	Stress must be above a certain threshold for SCC to occur. Limiting the rate of change in fluctuating stresses can significantly reduce the threshold stress.
Levels of CP	Adequate CP can prevent SCC except under disbonded coatings, which might shield the current from the pipe.

requirements and/or selecting design criteria such that the pipeline is not susceptible to SCC (such as using an FBE coating or operating at lower stress levels outside the susceptible levels for SCC) or controlling environmental factors, eliminating the corrosive environment that would lead to SCC/EIC (such as selecting backfill to control the pH level).

NACE SP0204 identifies some factors and rationale that may be considered to assist in mitigation. Table 7 provides some of these factors regarding susceptibility of SCC; selecting design and operating parameters accordingly helps to mitigate the potential for SCC.[12]

Monitoring activity is also important for SCC and can have significant O&M costs in the form of directly examining sections of piping that may be susceptible to cracking (e.g., selecting locations of highest stresses, pressure fluctuations, or temperatures and performing magnetic particle testing). Many operators deem it a best practice to inspect for SCC during any bellhole examination or direct inspection of pipelines because the incremental cost is minimal.

Mitigation Options for Cased Segments—Special Considerations

Cased pipeline segments were once believed to help improve integrity of pipelines, improving the safety to road and rail lines. Casings are now perceived to provide minimal benefit and have potential to be a significant external corrosion threat because shielding by the cased segment and/or electrolytic or metallic contact may strongly influence the safety effectiveness and negate the ability to monitor CP. A *Pipeline and Gas Journal* article summarizes a NACE panel interview that further details the rationale and trend to eliminate pipeline casings.[44]

Options to help assess and mitigate the threat of corrosion of a pipeline inside casings consist of the following:

- Eliminating the casing
- Filling the casing with a dielectric material (such as wax), provided the material stays intact
- Frequent monitoring (pipe-casing soil potential measurements, leak detection, performing ILI runs)

NACE SP0200[46] provides guidance for operators on managing cased pipelines.

The costs of mitigating, monitoring, and assessing casings have been extremely high because of practical difficulties and a shifting regulatory environment. Based on an American Gas Association (AGA)[(7)] survey presented at a 2010 U.S. (PHMSA) seminar, it is estimated that the majority of operators spend more than one third of their integrity management budget mitigating, assessing, and remediating cased segments, despite these segments' representing less than 2% of total pipeline mileage and virtually no incidents.

Mitigation Monitoring—Additional Sources of Information

Mitigation monitoring is an important step in the pipeline corrosion management process and confirms that mitigation strategies are operating as intended. Table 8 provides some basic techniques and relevant references on mitigation monitoring.

---

(7) American Gas Association (AGA), 400 N. Capitol St. NW, Suite 450, Washington, DC 20001.

TABLE 8 Additional Source on Mitigation Monitoring

Coupons	ANSI/NACE Standard RP0104, The Use of Coupons for Cathodic Protection Monitoring Applications[47]
Survey Techniques	NACE Standard TM0109, Aboveground Survey Techniques for the Evaluation of Underground Pipeline Coating Condition[48]
Close-Interval Survey (CIS) and Direct Current Voltage Gradient (DCVG) Surveys	NACE SP0207, Performing Close-Interval Potential Surveys and DC Surface Potential Gradient Surveys on Buried or Submerged Metallic Pipelines[49]
CP Measurement	NACE Standard TM0497, Measurement Techniques Related to Criteria for Cathodic Protection on Underground or Submerged Metallic Piping Systems[50]

Performing corrosion integrity assessments, covered in Section 7, is another important part of the pipeline corrosion management process also used to aid in the monitoring process and confirm that mitigation activities are functioning as intended.

SECTION 7

Integrity Assessments

The next step in the pipeline corrosion management process is to perform a condition assessment (also referred to as an integrity assessment), hereinafter referred to as "assessment," to confirm the integrity of the pipeline. The primary objective of an assessment is to identify and remediate areas on the pipeline where active corrosion exists or potential for future corrosion (e.g., areas of coating degradation) is more likely to occur, as well as determine whether any additional mitigation strategies are warranted. Answering the following questions is critical in determining the assessment approach and methodology:

- What are the historical and expected degradation mechanism(s) and threats of concern the assessment should detect?
- Are there existing inspection or other O&M data that can be leveraged in the assessment?
- What is the acceptance/failure criterion that will be used in the assessment of each threat?
- What is the optimal assessment approach that can answer these questions and what is the quality and type of information that will be collected?
- How will the assessment data be collected, transmitted, and stored?
- How will the data be disposed, who is the owner, and what is the priority and follow-up plan for anomalies?
- What is the remediation plan, and is there any feedback plan for greater mitigation activity?
- What are the costs associated with a particular assessment approach?

The flow chart in Figure 9 provides a basic process and framework for systematically answering the above questions for selecting and performing an assessment:

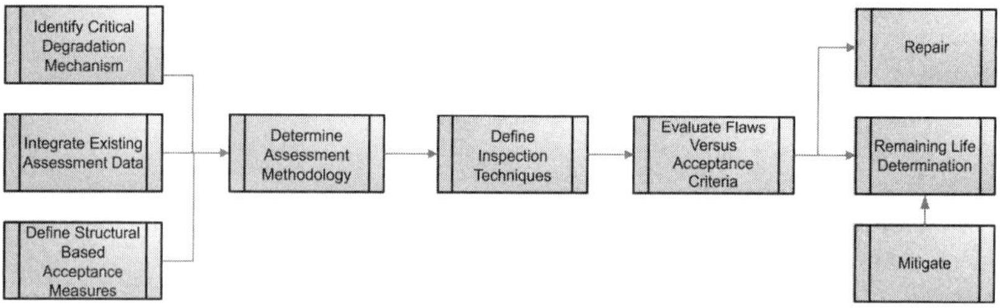

FIGURE 9 Example Assessment Process

There is a vast list of assessment tools and methodologies that can help validate the integrity of transmission pipelines along with new inspection technologies that are rapidly evolving to create new tools to better detect and estimate the size of anomalies. This section provides an overview and references for the following four basic categories of corrosion assessment tools:

- ILI or "smart pigging"
- Pressure testing (hydrotesting)
- Direct assessment (external corrosion direct assessment [ECDA], internal corrosion direct assessment [ICDA], stress corrosion cracking direct assessment [SCCDA])
- Other technologies (demonstrated to be equivalent to the other assessment methods)

In addition to the costs of performing an assessment, which can vary significantly among methodologies, the quality and type of information is critical in the decision-making process. Based on the process outlined in Figure 9, the selection of the methodology is best performed by evaluating the degradation mechanism (such as external corrosion) and the acceptance criteria (indication levels to identify the required sizing accuracy for which further action/inaction is needed). Once the data requirements of the assessment are identified, the capabilities along with costs can be evaluated. Previous assessments can also provide insight in terms of effectiveness of a particular approach and identify requirements for additional types of information that may be useful.

Data from a risk evaluation (including mitigation monitoring) may also be used to prioritize segments for assessment. Prioritizing high-risk pipeline segments yields a greater probability of detecting and remediating critical safety issues and results in the most cost-effective method for reducing the likelihood of a failure. Regulations in integrity management may influence the timing, type, and frequency of assessment. Tradeoffs also exist between the cost and the quality and/or thoroughness of the information collected for each integrity assessment. Evaluating the technical validity of the assessment, the thoroughness and accuracy of information collected, and the costs to complete an assessment on a particular segment is needed to determine the optimal choice.

Note that the cost may vary considerably even within the same assessment approach depending on the characteristics and location of a pipeline. For example, performing ECDA on a 10-mile (16-km) pipeline segment in a rural area over dirt soil with no drilling or traffic control will likely be considerably less costly than performing ECDA on a 5-mile (8-km) pipeline segment that runs along a busy freeway or dense urban area.

In-Line Inspection

ILI refers to the inspection of a pipeline segment from the interior of the pipe by having a specialized tool collect wall thickness (and crack) data as the tool is pushed by the flow down the pipeline. Different types of sensor technologies are incorporated in the tools depending on the type of corrosion threats believed to be present. Table 9 provides a brief description of some common tools and corrosion types that can be detected (more information can be found in API 1163[51] and NACE SP0102[52]).

Each ILI tool type and vendor technology has advantages and disadvantages as well as a range of cost implications that should be considered. These include

- Considerations regarding tool reliability and type of defects detected, ability to size those defects, and costs to run the tool.
- Capital expenditures/construction costs to add valves, bends, launchers, and receivers.
- Corresponding costs to cleaning the pipeline before the ILI tool is run.
- Operating expenditures for the inspection, operator oversight, and subsequent data analysis.
- Costs associated with the loss of throughput.
- Additional costs for directly examining a range of significant defects and anomalies as well as costs associated with recoating or other remediation.
- Not all lines were designed or able to be inspected via ILI; additional capital and constructions costs to make the line "piggable."
  - Some of these limitations are specified in API 1163 and NACE SP0102 and include tight bends, butterfly valves, select internal coatings, change in pipeline diameter, and others.

TABLE 9 ILI Tool Types and Characteristics

Type	Overview	Corrosion Defects Detected
Geometry Tool	An instrumented ILI tool that records data on localized deformation and geometry data about the pipeline internal circumference.	Ovality and Deformation Limited Corrosion Detection (Internal Only)
Magnetic Flux Leakage (MFL)	An ILI tool that induces an axial magnetic field in the pipe wall and analyzes changes in the distribution of magnetic field to identify a range of circumferential anomalies.	Sizing of External, Internal Corrosion
Transverse Magnetic Flux	An ILI tool similar in principle to MFL, but with a circumferential magnetic orientation to better detect axial anomalies.	Detection and Sizing of External and Internal Corrosion and SCC
Ultrasonic Testing (UT) Compression	An ILI tool that produces acoustic compression waves through transducer(s) and measures the normal beam acoustic response to detect defects. Typically requires a couplant.	Detection and Sizing of External and Internal Corrosion
UT Shear	An ILI tool that produces acoustic shear waves through transducer(s) and measures the acoustic response to detect defects. Typically requires a couplant.	Detection and Sizing of External and Internal Corrosion and SCC
Combined Tools	ILI tools that combine various methods above to provide multiple sets of data giving greater detail and information on pipeline defects.	Based on Combination of Choices

- Additional risk and cost factors; ILI vendors typically require purchase of tools if they are damaged during testing or extraction if stuck. Figure 10 shows a typical ILI tool.

ILI is one of the most accurate techniques to detect and size corrosion defects, although great variance in ILI costs may be expected depending on the application and requirements of the tool. Data mining and comparing to previous ILI runs can provide an even more complete picture of the integrity of the pipeline with the ability to trend the rate of corrosion by comparing changes in

FIGURE 10 Illustration of an ILI Tool

corrosion defects. Considerable work is required to verify sufficient data quality and make comparisons over time and for different vendor tool configurations. Table 10 lists sources that provide further information on ILI.

Hydrotesting

Hydrotesting is an assessment method that involves taking the pipe out of service and filling a pipeline segment with fluid (frequently water) and pressuring it to a level well above the maximum allowable operating pressure (MAOP).[54] Industry standards, regulations, and code requirements can help define the required test pressures. In some countries, hydrotesting is a regulatory requirement for all new pipelines and can be used to rerate or validate the MAOP.

One key advantage of hydrotesting is that it provides 100% inspection of the pipeline segment tested and verifies safe operation with the most significant defect in the pipeline.

Some disadvantages and cost and risk considerations that should be part of the evaluation of this assessment methodology include

- Interruption of service;
- Introduction of water and contaminants into the pipeline, which can result in corrosion damage;
- Challenges and costs associated with the purchase and environmentally safe disposal of fluid used for testing; and
- Provides only a snapshot in time confirming the pipeline is safe for operation at that instant and does not identify small defects or rate of deterioration of time-dependent threats.

TABLE 10 ILI References

ILI Standards	NACE SP0102, In-Line Inspection of Pipelines API 1163, In-Line Inspection Systems Qualification Standards ANSI/ASNT ILI-PQ, In-Line Inspection Personnel and Qualifications[53]

TABLE 11 Hydrotesting Reference

Hydrostatic Pressure Testing Reference	API RP 1110, Pressure Testing of Steel Pipelines for the Transportation of Gas, Petroleum Gas, Hazardous Liquids, Highly Volatile Liquids or Carbon Dioxide[55]

Costs primarily involve operating expenditures for performing the test with key benefits including high level of risk reduction and high confidence in ability to safely operate the pipeline. Industry estimates total costs on the order of four to seven times the costs of comparative assessment techniques.[57] Northern climates may require the use of preheated water and/or chemical mixtures (such as glycol or methanol) to prevent freezing. Table 11 lists a hydrotesting reference.

Direct Assessment

For pipelines on which it is not feasible or cost-effective to assess pipeline integrity using ILI or hydrotesting, DA may be a useful assessment tool. The objective of DA is to identify and prioritize those areas on a pipeline segment likely to exhibit corrosion and follow up by directly inspecting those areas over a scheduled period of time.

DA is a four-step process based on fundamental engineering practices that involves

(1) Preassessment: Collecting as much information about the design factors, construction, operation, and maintenance of the pipeline from SMEs, alignment sheets, design, and O&M records to determine the feasibility of DA and establish segmented regions for analysis.
(2) Indirect Inspection: Collecting field data and performing analysis of these data to supplement the preassessment data and prioritize areas likely to exhibit corrosion.
(3) Direct or Detailed Examination: Excavating and examining the pipeline at those identified areas as prioritized in Step 2. Usually repairs are also completed at this step to avoid duplicating costs/efforts even though they are officially considered part of postassessment (Step 4).
(4) Postassessment: Analyzing the results to determine whether repairs are required, determining the effectiveness of the approach, and identifying

TABLE 12 Direct Assessment Standards and References

ECDA	ANSI/NACE SP0502, Pipeline External Corrosion Direct Assessment Methodology[20] SP0507, External Corrosion Direct Assessment (ECDA) Integrity Data Exchange (IDX) Format[17] NACE SP0210, Pipeline External Corrosion Confirmatory Direct Assessment[23]
SCCDA	NACE SP0204, Stress Corrosion Cracking (SCC) Direct Assessment Methodology[12]
Dry Gas (DG)-ICDA	ANSI/NACE SP0206, Internal Corrosion Direct Assessment Methodology for Pipelines Carrying Normally Dry Natural Gas (DG-ICDA)[19]
Liquid Petroleum (LP)-ICDA	NACE SP0208, Internal Corrosion Direct Assessment Methodology for Liquid Petroleum Pipelines[21]
Wet Gas (WG)-ICDA	NACE SP0110, Wet Gas Internal Corrosion Direct Assessment Methodology for Pipelines[22]

future mitigation and remediation actions as well as a reassessment interval and steps for continuous improvement

DA is tailored for each particular corrosion threat with select unique data elements and inspection requirements based on the type of threat. NACE is the leader in identifying and standardizing processes to successfully perform DA for each corrosion threat with the sources of information in Table 12 providing more detailed information on performing DA.

Emerging Technologies

New technologies are constantly evolving and being introduced to the market to provide new and improved methods to assess transmission pipelines. In addition to the cost evaluation to procure and implement new technologies, evaluation should consider the consistency, thoroughness, and quality of information obtained.

Prior to these tools' receiving industry acceptance as an "other integrity assessment" technology, a certain amount of demonstration, testing, standards, and qualification expectations for those performing inspections must be developed. NACE has begun to write standards for guided wave testing (GWT);

TABLE 13 Emerging Technology and Assessment Tools

Technology	Overview	Applications
GWT: 	Guided wave testing uses piezoelectric transducers and other techniques to direct sound energy in wave modes, which creates guided waves within the pipe walls. These waves propagate down the pipe and reflect off features such as welds, supports, or areas of wall loss. The amplitude and symmetry of these reflections are then analyzed to determine the type and extent of these indications.	Aboveground segments, insulated piping, cased segments, short buried segments
Electromagnetic Acoustic Transducers: 	Couplant-free electromagnetic acoustic transducers (EMATs) can quickly screen components for corrosion, defects, gouges, and cracks, using electromagnetic induction to generate ultrasonic waves.	Inspection from the ID, larger-diameter pipe, aboveground piping/supports
Electromagnetic Wave (EMW) Inspection	EMW is a noninvasive, long-range inspection method to locate and identify anomalies and unfavorable conditions on cased and insulated pipelines. The system emits an electromagnetic signal and waits for a response reflected from objects; EMW equipment sends pulsed waves into the space surrounding a pipeline and receives the responses, which can then be located and identified.	Cased pipeline segments

Technology	Overview	Applications
Magnetic Tomography/Electric Induction Techniques	Pipeline inspection system that induces and analyzes magnetic fields on a pipeline from above ground. To create this magnetic field, an inspection current is induced between two contact points on the pipeline. Preferably these contact points are existing points on the pipe: e.g., CP test posts. Changes in magnetic field are then analyzed aboveground for changes or deviations to identify potential areas of wall loss.	TBD

others will be needed as technologies mature. The tools described in Table 13 are some relatively new technologies that can help assess pipeline segments in a limited capacity.

SECTION 8

Remediation

Remediation Overview

When faced with a pipe segment that has suffered damage from corrosion, an operator may choose to replace the segment, repair the segment, mitigate the corrosion process, or monitor and take no action at the time. Each of these strategies has multiple implementations. Choosing the correct path following an assessment or failure is a key component of safe, economical pipeline operation. These decision processes need to be outlined in the O&M manual before any assessment process is begun. The major issues that should be addressed as part of the remediation process with resultant decisions should be documented as follows:

- Does action need to be taken?
- How quickly and what type of actions need to be taken?
- Is a temporary solution necessary while preparations for the permanent remediation are being made?
- What are the regulatory and company policy requirements for action on this defect? What about the repaired pipeline?
- What type of remediation is most cost effective?
- Are qualified resources available?

Many remediation situations are time sensitive, especially those involving a significant risk of failure. Being prepared to make and execute the necessary decisions before the need arises helps expedite the process, thus increasing the safety and reducing the cost of pipeline operation.

Before proceeding with a remediation strategy, it is important for operators to understand the damage mechanism and limitations of the various remediation options. If corrosion damage is addressed improperly, then not only is remediation more costly but the likelihood of a failure can actually increase. Figure 11 presents a high-level overview of a remediation process.

Classification of Assessment Results

Ensuring valid assessment data through the use of quality control measures, such as comparison to known test points or features, is a critical step before any action is taken regarding remediation. Upon validation that the assessment results are providing sufficient and accurate data on the condition of the pipeline, classifying and identifying the severity of corrosion is critical in defining the appropriate course of action. As part of this classification, calculating the remaining strength of the pipeline by evaluating the burst pressure and safety factors as well as the estimated time to failure by reviewing corrosion rates can help determine the impact a corroded area has on pipeline integrity and the urgency of remediation activity.

Several different methods and analytical techniques are commonly used to assist in this evaluation, including ASME B31G,[57] modified ASME B31G, RSTRENG,[58] and log-secant (a methodology for evaluating crack-like defects) and are further reviewed in *The Pipeline Repair Manual*[59] by PRCI.[(8)]

Root-Cause Identification

Identifying the root cause of corrosion is a key step in selecting the correct remediation strategy. In addition, root-cause analysis provides additional insight to determine:

- Whether the assessment was suitable for finding the degradation caused by corrosion;
- The likelihood that similar corrosion damage will occur elsewhere on the pipeline;
- Whether the degradation is from a historic or active mechanism;
- Additional recommendations to mitigate this form of degradation on the pipeline; and
- The presence of interacting processes.

---

(8) Pipeline Research Council International (PRCI), 3141 Fairview Park Drive, Suite 525, Falls Church, VA USA 22042.

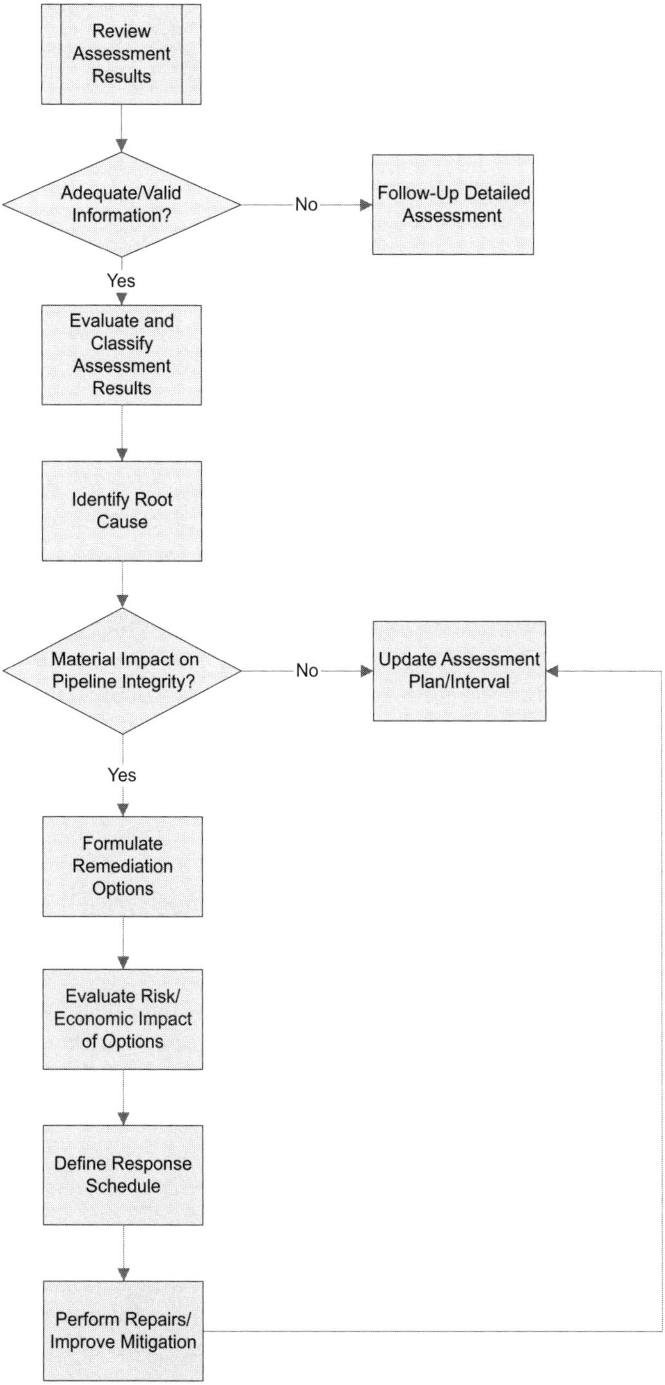

FIGURE 11 Pipeline Remediation Process

TABLE 14 Rehabilitation versus Replacement Costs

Pipeline Diameter	Comparison Cost of Rehabilitation to Cost of New Construction
150 to 350 mm (6 to 14 in)	70% to 130%
400 to 600 mm (16 to 24 in)	50% to 90%
700 to 1,200 mm (28 to 48 in)	25% to 50%

Pipe Remediation Options

Replacement

Replacement of a corroded pipeline segment is generally a last resort, to be used when the damage is widespread and severe. Two of the most important costs associated with a replacement project are interruption of service and emptying the line of hazardous substances. Costs from interruption of service are unique to each operator, depending on parallel lines, re-routing capabilities, and agreements with customers. If service cannot be interrupted, pipe just outside the edge of the planned replacement can be hot tapped. However, this further increases the expense of the replacement and may interfere with future ILI.

When a section of a gas pipeline is replaced, gas is typically vented to the atmosphere. The cost of the released gas can be the most expensive part of a replacement project.[60] Options for pumping down the line before the project to minimize the release of gas also exist. Liquid pipelines typically do not have the option to simply drain the line into the environment, and operators must arrange for the line to be drained safely.

Because of the replacement costs involved, pipeline repair may be more attractive than replacement as pipeline diameter increases. Table 14 provides ratios comparing rehabilitation to new construction as a function of pipeline diameter (presented at the International Pipeline Rehabilitation and Maintenance Conference, Prague, September 2000).[61]

Weld Deposition

When a localized area of corrosion is identified in a pipeline, it can be repaired by depositing weld metal directly into the corrosion pits to restore the original wall thickness. This type of repair is most economical for highly localized corrosion defects.

Weld repairs are usually made on the outside of a line but can remediate either external or internal corrosion damage.[62] The line can usually continue operating at full pressure while the repair is being conducted, but pressure is often reduced as a safety measure. Proper welding and nondestructive testing procedures should be used to ensure that the heating does not burn through or weaken the pipe to the point of failure, and that the heat-affected zone retains sufficient ductility. Hydrogen levels in the weld should be minimized by using a low-hydrogen electrode procedure and temper-bead deposition sequencing. Coating can then be applied over the completed and inspected repair.

When used to repair internal corrosion, a weld repair increases the safety factor of the line but does not affect the corrosion process. Internal corrosion repairs are temporary unless the root cause of the internal corrosion can be removed. Hence, weld repairs may also be used in conjunction with corrosion mitigation methods such as inhibitors, biocide, or line drying.

Weld deposition should not be used in instances in which remaining wall thickness is less than 3 mm (0.13 in) and should be prohibited near flash or ERW seams and crack-like defects.

Mechanical Sleeves

Steel sleeves can be added to a pipeline to reinforce the existing wall. This is the most common type of repair, with replacement a close second.[63] Steel sleeves are typically attached with long seam welds that do not impact the carrier pipe. Several steel sleeve designs can put the carrier pipe into compression with either bolts or heat. If a nonleaking sleeve is required, the ends are fillet welded in the girth direction.

A composite sleeve is also available, as are numerous types of composite wet wraps. Both types of composite repairs should be applied at atmospheric pressure after filler is first used to restore the original external pipe contour before the composite is wrapped. These composites are designed to minimize the strain bulging of the wall weakened by corrosion.

Sleeve repairs are very attractive when the pipe is easily accessed. Sleeves are typically effective, relatively inexpensive, and do not require any interruption in service. However, in areas such as near road crossings or river crossings, the pipe cannot be accessed directly to install a sleeve.

Pressure Reduction

Pressure reduction is typically used as a temporary measure when a near-critical defect is discovered or as a safety measure when personnel are working around defects that have not been directly measured. However, it can be used permanently if the impact on throughput is irrelevant or does not justify the costs of other repair methodologies. Generally the pressure is reduced to 80% of the earlier operating pressure, which represents the equivalent of a pressure test to 125% of the new pressure.

Crack Defects

A blunt defect has a higher safety factor than a crack with the same profile; thus cracks need to be removed. It is considered difficult to size the length and depth of cracks to assess their safety. When EAC is discovered, the preferred remediation method is to grind the cracks into corrosion-like defects. This is done in conjunction with magnetic particle testing because the cracks are usually not visible to the naked eye. Once grinding is complete, no crack remains, and the remaining strength calculations determine that the line is safe for operation, the line can recoated. If the safety factor of the blunt defect is unacceptable, the area should be repaired with a direct deposit weld repair or sleeve.

Coating Repairs

No buried pipeline system has a perfect coating. Holidays in the coating can be caused by handling during installation, rocks in the backfill, the coating application process, normal coating degradation, or other causes. CP is used to mitigate corrosion at these holidays. However, CP systems have limitations. As the coating degrades over time, disbonds, and the area of bare steel increases, more current is required by the CP system to maintain the required levels of polarization.

Recoating a pipeline improves the performance of the coating and CP system by reducing the area of bare steel that requires CP. A 2002 PRCI study found that coating repairs are effective for most types of coatings; however, it is important to choose the right type of repair for the main coating.[64] Numerous coating remediation methods are available, each with a unique set of advantages and disadvantages.[65]

Internal Linings

Linings may be installed to prevent further internal corrosion on a pipeline by creating a barrier between the steel and electrolyte. The pipeline must be taken out of service and cleaned to help proper adhesion. It is usually impractical to apply CP to the inside of the pipeline, so that any areas that are not coated (welds, holidays) may experience accelerated corrosion. One of the major advantages of this method is that it is not necessary to excavate the entire line, as significant lengths of pipeline can be lined from a single access point. However, this lining may be damaged with maintenance pigging and/or interfere with ILI tools by increasing lift-off, and may be a structural impediment to tool travel through the repaired section. In addition, sudden decompression of the line can cause the liner to collapse if products (such as methane) diffuse over time and become trapped between the steel and the repair layer.

Internal linings can add strength to a pipeline.[66] Installation of this type of repair poses technical challenges but may be an attractive alternative to replacement in areas where excavation of the whole segment is impossible or time is required to make a more permanent repair.

Cathodic Protection System Remediation

Improvement of a buried pipeline's CP system is considered to be a type of remediation for external corrosion. As long as a pipe with external corrosion has an acceptable safety factor (failure pressure/operating pressure), mitigating any additional corrosion damage is an attractive way to address the deterioration problem. Typical CP remediation activities include

- Installation of additional anode beds and/or rectifiers;
- Installation of a galvanic anode to locally fortify an impressed current system;
- Increasing the output of an impressed current system until protection criteria are met; and
- Adding bonds to other buried structures to address stray current issues.

It is important to note that impressed current systems are subject to lightning strikes, vandalism, electrical maintenance, and anode replacement, as galvanic and impressed current deplete over time. Thus, it is important to properly monitor and maintain a CP system.

CP improvements can address only certain problems. For instance, it is possible for external corrosion to occur when coatings disbond and electrically shield, preventing CP current from reaching the pipe. Improving the CP system in this case does not protect the system. This highlights the importance of fully understanding the cause of a corrosion problem before taking action.

CP systems are discussed in detail in Section 6.

TABLE 15 Remediation Example Response Tiers

Response Tier	Criteria	Response Time
Immediate	Failure Pressure <110% MAOP	5 days
Scheduled	Failure Pressure >110% MAOP Will grow to critical dimensions before next scheduled inspection	0 to 20 years, depending on expected corrosion rate, failure pressure, and operating stress level
Monitored	Will NOT grow to critical dimensions before next scheduled inspection	Reevaluate at next scheduled inspection

Response Schedule

Assessments often find multiple indications of corrosion throughout the pipeline system. Some defects remain relatively dormant for years while others require a more immediate response. Operators should develop a method for prioritizing and scheduling required responses to a range of expected defects so that resources can be properly allocated when defects are discovered. ASME B31.8S recommends a prioritized response schedule based on a three-tiered response approach as outlined in Table 15.[10]

Operators may wish to include information about previous inspections, consequences of failure information, and rate of degradation in determining the priority in the response schedule as well as other guidelines available from industry associations such as NACE, PRCI, API, and ASME.

Economic and Risk Considerations

A corrosion defect can be addressed in several ways: the defect may require repair, and additional remedial measures may be applied to mitigate further corrosion. The best option is the one that maximizes the reliability and profits from the line while complying with applicable regulations and achieving an acceptable level of safety. Many of the costs are incurred in the future; to correctly compare these with present costs, see the discussion of *net present value* (NPV) in Appendix A.

Safety Considerations

Because of the high-risk nature of certain pipeline segments, operators may wish to define minimum criteria based on perceived risk to public safety or property.

Cost of Failure

Unlike planned costs such as capital or maintenance expenditures, a failure is a risk-based event that may or may not occur. However, the expected cost of a failure may be directly compared with fixed costs. Although neither probability nor expected cost can be calculated exactly, estimates can be derived from previous accidents, industry statistics, and other methods.

The cost of failure justifies several options in addition to or in lieu of remediation. These options include

- Take no action and deal with failures after they occur;
- Slow or eliminate degradation through additional mitigation activity such as CP;
- Take the line out of service or reduce the MOP to increase the safety factor; or
- Remediate or repair the identified issue.

Cost of Further Degradation

Another important consideration in the remediation process is the nature of degradation and the distribution of the corrosion rate if known. The expected corrosion rate provides input into the remaining life of the pipeline but has considerable uncertainty. Ideally the rate of corrosion following remediation should return to a low value. In the event in which the expected corrosion rate remains significant, it is likely that a repair or replacement may be needed in the future. The cost of such a future repair should be considered in the economic analysis and should take into account the expected corrosion rate.

Value of Information

Information is particularly important in the remediation process. Statistics such as the corrosion rate or precise dimensions of a flaw are often available for a cost but can provide invaluable information in determining the appropriate course of action.

Expected Life of Pipeline

Properly maintained, a pipeline can be operated for extended periods of time. However, the transportation requirements of a pipeline often change over time. Wells can be depleted, leaving the line with little or nothing to transport. Demand for gas or crude in an area can increase, requiring a pipeline with greater capacity. If a pipeline requires repairs but has a limited lifetime, sometimes a cheaper, temporary solution is preferable over a more expensive, permanent solution.

Regulatory Requirements

Operational/Material Requirements

As regulations evolve, newly laid pipelines are often held to a higher corrosion prevention standard than in-service pipelines. This may cause a replaced pipeline segment to have different regulatory requirements for materials, coating, CP, frequency of inspection, type of inspection, and other factors. This is an important part of the economic analysis during consideration of repair versus replacement of a pipeline segment.

Defect Requirements

Some corrosion defects are required to be repaired upon discovery. Applicable regulations and industry standards should be reviewed and incorporated into the remediation decision process.

Key Resources

PRCI Pipeline Repair Manual

The most recent Pipeline Repair Manual (PRM) was developed for PRCI in 2006. It provides details on the analysis of defects and selection of repair methods. It is available from PRCI: http://prci.org.

SECTION 9

Program Effectiveness

Effectiveness Monitoring and Feedback into the Corrosion Management Process

Each pipeline operator has unique challenges. Pipelines vary by physical environment, consequence of failure, regulatory requirements, and many other variables. This guide contains guidelines for implementing a corrosion management process, but the specifics depend on the operator that implements it. Continually adapting the program to better reflect the needs of the company and pipeline system increases effectiveness and helps an operator to correctly allocate resources.

For instance, an operator may find during an annual review that a large portion of the corrosion management budget is going toward repairing external corrosion damage. The operator may want to consider improving the mitigation system to prevent this damage from taking place. An alternative would be pursuing more cost-effective repair methods or replacement. If the improvements are successful, the pipeline operates more safely and at the same or a possibly reduced cost.

Inputs to a Continuous Improvement Process

Individual perceptions and anecdotal evidence make good starting points for gathering data about a particular aspect of the program. However, such perceptions do not give a complete picture of the program. Focusing continuous improvement efforts on gathering factual information creates a more complete picture of what is happening and helps to generate consensus. Performance metrics that an operator may want to track include

- Number of leaks, amount of product released, average cleanup cost
- Number of repairs, type of repair, urgency of repairs

- Number and length of service interruptions
- Incidents caused by corrosion failure
- CP potentials, current density, and trends
- Internal corrosion coupon corrosion rates
- Assessment costs by type, per mile (km) of pipeline, per mile (km) of high-consequence area (HCA)

Compliance with Plan

Corrosion management plans are often complex and written in advance of being implemented. One reason for poor plan performance can be that some individuals or parts of an organization do not comply. It may be helpful to track the number of activities that are properly documented, and of those that are, how many were conducted in compliance with the plan. If compliance is a problem, adjustments should be identified and a communications plan should include a process to ensure that the corresponding adjustments are implemented.

Goals of Effectiveness Monitoring

Increase Efficiency

Studying performance metrics and continually improving the process can eliminate waste, provide better protection, and/or increase effectiveness of the program. Allocating budget to activities such as mitigation, assessment, and remediation is complicated by the fact that the success of one aspect reduces the need for others. For instance, if a high-quality assessment discovers corrosion protection system problems early, the problems can be addressed without a costly replacement. Each operator needs to find the correct balance based on physical environment, regulatory constraints, and other factors.

Improve Safety

Safety of the public as well as the operator's employees is of paramount importance. Properly managing corrosion can significantly improve the safety and reliability of pipeline transportation.

Set Goals for the Program and Measure Outcomes

Managers can set goals based on performance measures and use the continuous improvement process to track progress toward those goals. Goals may change over time to ensure best use of the resources.

Possible Strategies

Some suggestions for continuous improvement programs are presented here.

Database Tools and Software

Database tools and specialized software programs may be used to track, trend, and interrelate corrosion-related data. A line segment can be related in database software to its coating type, CP system, material type, welding procedures, previous repairs, or any other pertinent information. If the database is properly maintained, an operator can measure change over time, determine operational correlations, and observe the relationship between plan inputs, such as budget dollars or assessment strategy, and performance results such as the number of repairs.

Annual Program Review

Many operators conduct an annual program review, which presents an opportunity for individuals to present their new ideas and perceptions to improve the program. These ideas can be cross-referenced, compared, and validated with available data.

Geographic Information Systems

GIS applications are designed to display spatial relations and can be updated to display relationships between O&M investments, capital spending, performance, and other factors such as terrain or population density. The output of the application is typically a map displaying the spatial relationship between variables. Many operators have implemented GIS to some extent, but the use of these systems could be improved by better integrating with corrosion-related data. For example, GIS could display observed coating damage incidents overlaid on a geological map of soil types to determine whether there is a useful correlation. In this case, the operator may want to identify and use a more damage-resistant coating in areas where the soil is particularly aggressive, causing stretching and disbonding of coatings.

References

1. International Energy Agency, "Share of total primary energy supply in 2007," Energy supply by fuel type, 2009, http://www.iea.org/stats/pdf_graphs/29TPESPI.pdf (Oct. 1, 2009).

2. M. Mohitpour, *Energy Supply and Pipeline Transportation Challenges and Opportunities* (New York, NY: ASME, 2008), p. 109.
3. U.S. Department of Transportation, "Significant Pipeline Incidents by Cause," U.S. Pipeline Incidents by Cause, http://primis.phmsa.dot.gov/comm/reports/safety/PSI.html.
4. N.G. Thompson, "Cost of Corrosion Control, Appendix E: Gas and Liquid Transmission Pipelines," U.S. Federal Highway Administration (FHWA), 2002, http://www.corrosioncost.com/pdf/gasliquid.pdf (Oct. 1, 2010).
5. G.A. Jacobson, "Corrosion 101, A Fundamental Overview of Corrosion," NACE, http://events.nace.org/library/articles/corrosion101.asp (Oct. 1, 2010).
6. M. Baker Jr., "Integrity Management Program Delivery Order DTRS56-02-D-70036 Stress Corrosion Cracking Study Final Report," Department of Transportation, http://primis.phmsa.dot.gov/iim/docstr/SCC_Report-Final_Report_with_Database.pdf, January 2005, pp. 15-22 (Oct. 1, 2010).
7. A. deS. Brasunas, *Corrosion Basics: An Introduction* (Houston, TX: NACE, 1984).
8. NACE Publication 35103 (latest revision), "External Stress Corrosion Cracking of Underground Pipelines" (Houston, TX: NACE).
9. R. Fessler, M. Baker Jr., "Pipeline Corrosion Final Report," U.S. Department of Transportation PHMSADTRS56-02-D-70036, November 2008, pp. 26–29.
10. ANSI/ASME B31.8S (latest revision), "Managing System Integrity of Gas Pipelines" (New York, NY: ASME).
11. ANSI/NACE MR0175/ISO 15156, "Petroleum and natural gas industries—Materials for use in H_2S-containing environments in oil and gas production" (Houston, TX: NACE).
12. NACE SP0204 (latest revision), "Stress Corrosion Cracking (SCC) Direct Assessment Methodology" (Houston, TX: NACE).
13. U.S. Code of Federal Regulations (CFR), Title 49, "Protection Against Accidental Overpressure, Part 192 (Washington, DC: Office of the Federal Register, 1995).
14. U.S. Code of Federal Regulations (CFR), Title 49, "Protection Against Accidental Overpressure," Part 195 (Washington, DC: Office of the Federal Register, 1995).
15. API 1160 (latest revision), "Managing System Integrity for Hazardous Liquid Pipelines" (Washington, DC: API).
16. NACE Publication 3C194 (withdrawn), "Economics of Corrosion" (Houston, TX: NACE). Available from NACE as a historical document only.
17. NACE SP0507 (latest revision), External Corrosion Direct Assessment (ECDA) Integrity Data Exchange (IDX) Format (Houston, TX: NACE).
18. W.K. Muhlbauer, *Pipeline Risk Management Manual*, 2nd ed. (Houston, TX: Gulf Professional Publishing, 1999), pp. 13–16.

19. NACE SP0206 (latest revision), "Internal Corrosion Direct Assessment Methodology for Pipelines Carrying Normally Dry Natural Gas (DG-ICDA)" (Houston, TX: NACE).

20. NACE SP0502 (latest revision), "Pipeline External Corrosion Direct Assessment Methodology" (Houston, TX: NACE).

21. NACE SP0208 (latest revision), "Internal Corrosion Direct Assessment Methodology for Liquid Petroleum Pipelines" (Houston, TX: NACE).

22. NACE SP0110 (latest revision), "Wet Gas Internal Corrosion Direct Assessment Methodology for Pipelines" (Houston, TX: NACE).

23. NACE SP0210 (latest revision), "Pipeline External Corrosion Confirmatory Direct Assessment" (Houston, TX: NACE).

24. W.K. Muhlbauer, *Pipeline Risk Management Manual*, 2nd ed. (Houston, TX: Gulf Professional Publishing, 1999), pp. 24–26.

25. NACE SP0169 (latest revision), "Control of External Corrosion on Underground or Submerged Metallic Piping Systems" (Houston, TX: NACE).

26. C. Munger, L. Vincent, *Corrosion Prevention by Protective Coatings,* 2nd ed. (Houston, TX: NACE, 1999), pp. 300–308.

27. NACE No. 1/SSPC-SP 5 (latest revision), "White Metal Blast Cleaning" (Houston, TX: NACE, and Pittsburgh, PA: SSPC).

28. ISO 8501-1 (latest revision), "Preparation of steel substrates before application of paints and related products—Visual assessment of surface cleanliness—Part 1: Rust grades and preparation grades of uncoated steel substrates and of steel substrates after overall removal of previous coatings" (Geneva, Switzerland: ISO).

29. NACE No. 2/SSPC-SP 10 (latest revision), "Near-White Metal Blast Cleaning" (Houston, TX: NACE, and Pittsburgh, PA: SSPC).

30. NACE No. 3/SSPC-SP 6 (latest revision), "Commercial Blast Cleaning" (Houston, TX: NACE, and Pittsburgh, PA: SSPC).

31. SSPC-SP 8 (latest revision), "Pickling" (Pittsburgh, PA: SSPC).

32. NACE No. 4/SSPC-SP 7 (latest revision), "Brush-Off Blast Cleaning" (Houston, TX: NACE, and Pittsburgh, PA: SSPC).

33. SSPC-SP 3 (latest revision), "Power Tool Cleaning" (Pittsburgh, PA: SSPC).

34. SSPC-SP 2 (latest revision), "Hand Tool Cleaning" (Pittsburgh, PA: SSPC).

35. SSPC-SP 1 (latest revision), "Solvent Cleaning" (Pittsburgh, PA: SSPC).

36. NACE No. 5/SSPC-SP 12 (latest revision) "Surface Preparation and Cleaning of Metals by Waterjetting Prior to Recoating" (Houston, TX: NACE, and Pittsburgh, PA: SSPC).

37. NACE No. 8/SSPC-SP 14, (latest revision) "Industrial Blast Cleaning" (Houston, TX: NACE, and Pittsburgh, PA: SSPC).

38. A.W. Peabody, R.L. Bianchetti, eds., *Peabody's Control of Pipeline Corrosion,* 2nd ed. (Houston, TX: NACE, 2001), p. 18.

39. R. Fessler, M. Baker Jr., "Pipeline Corrosion Final Report," U.S. Department of Transportation Pipeline and Hazardous Materials Safety Administration, DTRS56-02-D-70036, November 2008, p. 17.

40. A.W. Peabody, R.L. Bianchetti, eds., *Peabody's Control of Pipeline Corrosion,* 2nd ed. (Houston, TX: NACE, 2001), p. 21.

41. A.W. Peabody, R.L. Bianchetti, eds., *Peabody's Control of Pipeline Corrosion,* 2nd ed., (Houston, TX: NACE, 2001), p. 23.

42. A.W. Peabody, R.L. Bianchetti, eds., *Peabody's Control of Pipeline Corrosion,* 2nd ed., (Houston, TX: NACE, 2001), p. 33.

43. ANSI/NACE SP0106 (latest revision), "Control of Internal Corrosion in Steel Pipelines and Piping Systems" (Houston, TX: NACE).

44. "Corrosion Inhibitor," Wikipedia, September 1, 2010, http://en.wikipedia.org/wiki/Corrosion_inhibitor (Oct. 1, 2010).

45. "Protecting Pipelines At Crossings: Are Casings Obsolete?", Special to *Pipeline & Gas Journal* 236, 3 (available at the following link): (http://pipelineandgasjournal.com/protecting-pipelines-crossings-are-casings-obsolete?page=show).

46. NACE SP0200 (latest revision), "Steel-Cased Pipeline Practices" (Houston, TX: NACE).

47. ANSI/NACE Standard RP0104 (latest revision), "The Use of Coupons for Cathodic Protection Monitoring Applications" (Houston, TX: NACE).

48. NACE Standard TM0109 (latest revision), "Aboveground Survey Techniques for the Evaluation of Underground Pipeline Coating Condition" (Houston, TX: NACE).

49. NACE SP0207 (latest revision), "Performing Close-Interval Potential Surveys and DC Surface Potential Gradient Surveys on Buried or Submerged Metallic Pipelines" (Houston, TX: NACE).

50. NACE Standard TM0497 (latest revision), "Measurement Techniques Related to Criteria for Cathodic Protection on Underground or Submerged Metallic Piping Systems" (Houston, TX: NACE).

51. API 1163 (latest revision), "In-Line Inspection Systems Qualification Standard" (Washington, DC: API).

52. NACE SP0102 (latest revision), "In-Line Inspection of Pipelines" (Houston, TX: NACE).

53. ASNT ILI-PQ (latest revision), "In-Line Inspection Personnel and Qualifications" (Columbus, OH: ASNT).

54. R. Fessler, M. Baker Jr., "Pipeline Corrosion Final Report," U.S. Department of Transportation PHMSA DTRS56-02-D-70036, November 2008, p. 35.

55. N.G. Thompson, "Cost of Corrosion Control, Appendix E: Gas and Liquid Transmission Pipelines," U.S. Federal Highway Administration, 2002, http://www.corrosioncost.com/pdf/gasliquid.pdf, p. E36 (Oct. 1, 2010).

56. API RP 1110 (latest revision), "Pressure Testing of Steel Pipelines for the Transportation of Gas, Petroleum Gas, Hazardous Liquids, Highly Volatile Liquids or Carbon Dioxide" (Washington, DC: API).

57. ANSI/ASME B31G (latest revision), "Determining the Remaining Strength of Corroded Pipelines" (New York, NY: ASME).

58. Technical Toolboxes, "RSTRENG," http://www.ttoolboxes.com/products/rstreng.htm (Nov 4, 2010).

59. C.E. Jaske, B.O. Hart, W.A. Bruce, "Pipeline Repair Manual," PRCI Project No. PR-186-0324, (Houston, TX: Technical Toolboxes, 2006), pp. 3–4.

60. J. Baek, et. al., "Comparative Study for Various Repair Methods of In-Service Pipeline Using Full Scale Burst Test," 23rd World Gas Conference, Amsterdam, The Netherlands, 2006.

61. R. Palmer-Jones, D. Paisley, "Repairing Internal Corrosion Defects in Pipelines – A Case Study," 2006 http://www.penspenintegrity.com/downloads/RepairInternalCorrosion.pdf, (Oct. 1, 2010), p. 5.

62. Y. Wang, W.A. Bruce, "Examination of External Weld Deposition Repair for Internal Wall Loss," Pipeline Research Council International, Project No. PR-185-9633 (Arlington, VA: PRCI, March 1998).

63. I. Harris, "Internal Repair of Gas Pipelines Survey of Operator Experience and Industry Needs Report," Office of Scientific and Technical Information, 2003, http://www.osti.gov/ (Oct. 1, 2010).

64. N.G. Thompson, J.A. Beavers, G. Ruschau, "Compatibility of Repair Coatings to Existing Below-Grade Coatings," PRCI Project No. PR-186-0029 (Arlington, VA: PRCI, April 2002).

65. NACE Publication 10D199 (withdrawn), "Coatings for the Repair and Rehabilitation of the External Coatings of Buried Steel Pipelines" (Houston, TX: NACE). Available from NACE as a historical document only.

66. "Internal Repair of Pipelines – Final Technical Report," PRCI Project No. PR-185-0367 (Arlington, VA: PRCI, December 2005).

67. R. Brealy, S. Myers, *Principles of Corporate Finance* (New York, NY: McGraw-Hill Irwin, 2003), p. 14.

68. "Corrosion-Economics," http://events.nace.org/library/corrosion/Economics/Present.asp.

69. NACE Standard RP0394 (latest revision), "Application, Performance, and Quality Control of Plant-Applied, Fusion-Bonded Epoxy External Pipe Coating" (Houston, TX: NACE).

70. NACE Standard RP0402 (latest revision), "Field-Applied Fusion-Bonded Epoxy (FBE) Pipe Coating Systems for Girth Weld Joints: Application, Performance, and Quality Control" (Houston, TX: NACE).

71. API RP 5L7 (latest revision), "Unprimed Internal Fusion Bonded Epoxy Coating of Line Pipe" (Washington, DC: API).

72. CSA Z245.20-M92 (latest revision), "External Fusion Bonded Epoxy Coating for Steel Pipe" (Ontario, Canada: CSA).

73. ANSI/AWWA C 213 (latest revision), "Fusion-Bonded Epoxy Coating for the Interior and Exterior of Steel Water Pipelines" (Denver, CO: AWWA).

74. ISO 21809-2 (latest revision), External coatings for buried or submerged pipelines used in pipeline transportation systems—Part 2: Fusion-bonded epoxy coatings (Geneva, Switzerland: ISO).

75. R. Fessler, M.l. Baker Jr., "Pipeline Corrosion Final Report," U.S. Department of Transportation PHMSA DTRS56-02-D-70036, November 2008, p. 16.

76. NACE Standard RP0105 (latest revision), "Liquid-Epoxy Coatings for External Repair, Rehabilitation, and Weld Joints on Buried Steel Pipelines" (Houston, TX: NACE).

77. ISO 21809-3 (latest revision), External coatings for buried or submerged pipelines used in pipeline transportation systems—Part 3: Field-joint coatings (Geneva, Switzerland: ISO).

78. NACE SP0185 (latest revision), "Extruded Polyolefin Resin Coating Systems with Soft Adhesives for Underground or Submerged Pipe" (Houston, TX: NACE).

79. NACE Standard RP0399 (latest revision), "Plant-Applied External Coal Tar Enamel Pipe Coating Systems: Application, Performance, and Quality Control" (Houston, TX: NACE).

80. ISO 21809-4 (latest revision), External coatings for buried or submerged pipelines used in pipeline transportation systems—Part 4: Polyethylene coatings (Geneva, Switzerland: ISO).

81. NACE Standard RP0602 (latest revision), "Field-Applied Coal Tar Enamel Pipe Coating Systems: Application, Performance, and Quality Control" (Houston, TX: NACE).

82. ANSI/AWWA C 203 (latest revision), "AWWA Standard for Coal-Tar Protective Coatings and Linings for Steel Water Pipelines—Enamel and Tape—Hot Applied" (Denver, CO: AWWA).

83. SSPC-Paint 32 (latest revision), "Coal Tar Emulsion Coating" (Pittsburgh, PA: SSPC).

84. SSPC-Paint 33 (latest revision), "Coal Tar Mastic, Cold Applied" (Pittsburgh, PA: SSPC).

85. SSPC-Paint 16 (latest revision), "Coal Tar Epoxy Black (or Dark Red) Paint" (Pittsburgh, PA: SSPC).

86. CSA Z245.21-M92 (latest revision), "External Polyethylene Coating for Pipe" (Ontario, Canada: CSA).

87. ANSI/AWWA C214 (latest revision), "Standard for Tape-Coating Systems for the Exterior of Steel Water Pipelines" (Denver, CO: AWWA).

88. NACE Standard RP0303 (latest revision), "Field-Applied Heat-Shrinkable Sleeves for Pipelines: Application, Performance, and Quality Control" (Houston, TX: NACE).

89. NACE Standard RP0375 (latest revision), "Field-Applied Underground Wax Coating Systems for Underground Pipelines: Application, Performance, and Quality Control" (Houston, TX: NACE).

90. ISO 21809-3 (latest revision), "External coatings for buried or submerged pipelines used in pipeline transportation systems—Part 3: Field joint coatings" (Geneva, Switzerland, ISO).

91. ISO 8504-3 (latest revision), "Preparation of steel substrates before application of paints and related products—Surface preparation methods—Part 3: Hand- and power-tool cleaning" (Geneva, Switzerland: ISO).

92. SSPC-Paint 34 (latest revision), "Water-Borne Epoxy Topcoat for Steel Surfaces" (Pittsburgh, PA: SSPC).

93. ANSI/AWWA C210 (latest revision), "Standard for Liquid-Epoxy Coating Systems for the Interior and Exterior of Steel Water Pipelines" (Denver, CO: AWWA).

94. SSPC-PS 4.04, "Four-Coat White or Colored Vinyl Painting System" (Pittsburgh, PA: SSPC).

95. SSPC-PS 4.02, "Four-Coat Vinyl Painting System" (Pittsburgh, PA: SSPC).

96. SSPC Guide 4.00, "Guide for Selecting Vinyl Painting Systems" (Pittsburgh, PA: SSPC).

97. SSPC-Paint 27 (latest revision), "Basic Zinc Chromate-Vinyl Butyral Wash Primer" (Pittsburgh, PA: SSPC).

98. SSPC-Paint 9 (latest revision), "White (or Colored) Vinyl Paint" (Pittsburgh, PA: SSPC).

99. SSPC-Paint 8 (latest revision), "Aluminum Vinyl Paint" (Pittsburgh, PA: SSPC).

100. SSPC-Paint 106 (latest revision), Black Vinyl Paint (Pittsburgh, PA: SSPC).

101. SSPC-Paint 30 (latest revision), "Weld-Through Inorganic Zinc Primer" (Pittsburgh, PA: SSPC).

102. SSPC-Paint 20 (latest revision), "Zinc-Rich Coating (Type 1 - Inorganic, and Type II – Organic" (Pittsburgh, PA: SSPC).

103. SSPC-Guide 14, "Guide for the Repair of Imperfections in Galvanized, Organic, or Inorganic Zinc-Coated Steel Using Organic Zinc-Rich Coatings" (Pittsburgh, PA: SSPC).

104. SSPC-Paint 29 (latest revision), "Zinc Dust Sacrificial Primer, Performance-Based" (Pittsburgh, PA: SSPC).

105. SSPC PS Guide 12.00, "Guide to Zinc-Rich Coating Systems" (Pittsburgh, PA: SSPC).

106. SSPC-PS Guide 8.00, "Guide to Topcoating Zinc-Rich Primers" (Pittsburgh, PA: SSPC).

107. SSPC-PS 15.04, "Chlorinated Rubber Painting System for Field Application over a Shop-Applied Solvent" (Pittsburgh, PA: SSPC).

108. SSPC-Guide 20.00, "Guide for Selecting Painting Systems for Boottoppings" (Pittsburgh, PA: SSPC).

109. NACE No. 12/SSPC-CS23.00/AWS C2.23M (latest revision), "Specification for the Application of Thermal Spray Coatings (Metallizing) of Aluminum, Zinc, and Their Alloys and Composites for the Corrosion Protection of Steel" (Houston, TX: NACE).

110. ANSI/AWWA C222 (latest revision), "Standard for Polyurethane Coatings for the Interior and Exterior of Steel Water Pipe and Fittings" (Denver, CO: AWWA).

111. Norsok M501 (latest revision), "Surface preparation and protective coating" (Lysaker, Norway: Standards Norway).[9]

112. ASTM D5894 (latest revision), "Standard Practice for Cyclic Salt Fog/UV Exposure of Painted Metal (Alternating Exposures in a Fog/Dry Cabinet and a UV/Condensation Cabinet" (West Conshohocken, PA: ASTM).

113. ISO 12944 (latest revision), "Paints and varnishes—Corrosion protection of steel structures by protective paint systems" (Geneva, Switzerland, ISO).

114. "Cost Benefit Analysis," Wikipedia: http://en.wikipedia.org/wiki/Cost-benefit_analysis (October 1, 2010).

[9] Standards Norway, P.O. Box 252, NO-1326, Lysaker, Norway.

APPENDIX A

Net Present Value Explanation and Example

Managers are often challenged with how to make decisions regarding capital investments and whether to pursue certain projects. NPV is one of the most important economic analysis tools that can be used to evaluate the value of these projects because it accounts for the time value of money, or the concept that a dollar today is worth more than a dollar tomorrow because of the ability to invest and earn compounded interest.[68]

An NPV analysis can be very useful in evaluating today's normalized cost of certain corrosion mitigation, assessment, and remediation projects, when the current and future costs and benefits are quantified into a series of cash flows over time. These cash flows can then be discounted at a firm's designated discount or hurdle rate to determine the overall value of the project. Positive NPV projects are viewed as accretive to company value and may be deemed worth pursuing.

The following equations may be used to calculate NPV.[68]

The NPV is a summation of the present value of all cash inflows and outflows minus the initial project cost (C_0). To include the effects of taxation (essentially a business expense), all actual cash flows for tax-paying organizations are reduced by the formula given by Equation (A1):

$$\text{amount after taxes} = C \cdot (1 - T_x) \tag{A1}$$

Depreciation allowances are an excellent example of when such tax savings are possible; they are treated similarly to income. All expenses allowed to be charged against income for tax purposes, but not representing actual cash flow, are modified by Equation (A2):

$$\text{Cash flow} = \text{non-cash expense charge } (T_x) \tag{A2}$$

The present value (PV) of the tax savings cash flow from a depreciation expense series (DES) of an original cost is given by Equation (A3):

$$PV_{DES} = C_0(Q, i, N_Q) \cdot T_x \tag{A3}$$

- where Q is the present value factor for a cash flow stream and
- N_Q is the time span of the depreciation expense stream

Combining these definitions, one can obtain the fundamental NPV Equation (A4) that includes the tax effects:

$$NPV = \sum_{n=1}^{N} \frac{F_n}{(1+i)^n}(1 - T_x) - C_0(Q, i, N_Q) \cdot T_x - C_0 \tag{A4}$$

Examples 1 and 2 provide a basic qualitative analysis and NPV-based analysis for making decisions with regard to pipeline corrosion and investment decisions:

Example 1: Qualitative Analysis

A municipality requested a CP design for sheet pilings for a drainage canal. The consultant proposed two designs, an impressed current system with bare pilings, because much of the coating would be damaged when the piles were driven, and a galvanic system, which required coated pilings to reduce the current demand to an achievable level and had a much higher initial cost. Ultimately the municipality engineers selected the galvanic system with coating because of the lack of availability of skilled personnel at the municipality to operate and maintain the impressed-current CP system.

Example 2: Quantitative—NPV-Based Analysis

A pipeline operator has identified a 5-mile (8-km) segment that is experiencing corrosion, and the operator is evaluating a decision to recoat and extend the life for 10 years or replace the pipeline in 3 years. Replacing the line would enable an increased throughput, increasing the cash generation of the asset by 20%.

Project Assumptions:

Option One: Pipeline Recoat

- Costs associated with recoating: $2,000,000
- Asset revenue generation (annual): $500,000
- Loss of throughput during recoat: $50,000

Option Two: Pipeline Replacement

- Costs associated with replacement $5,000,000
- Asset revenue generation (with 20% increase for product throughput): $600,000
- Asset life (post-replacement): 12 Years

Company-Specific Assumptions Applicable to Both Options:

- For simplicity, assume no impact from taxes (i.e., tax rate: 0%)
- Weighted cost of capital (i): 11%
- Expected remaining life of the asset is 10 years after recoating
- Expected remaining life of the asset is 12 years after replacement

Table A1 shows the net cash flows and corresponding NPV associated with each option. As evident from the example, option one has the higher value based on the NPV calculation.

TABLE A.1 Series of Cash Flows

Option One: Recoat	Year 0	Year 1	Year 2	Year 3	Year 4	Year 5	Year 6	Year 7	Year 8	Year 9	Year 10	Year 11	Year 12	Year 13	Year 14	Year 15
Asset Cash Flow Generation	$450,000	$500,000	$500,000	$500,000	$500,000	$500,000	$500,000	$500,000	$500,000	$500,000	$500,000	$0	$0	$0	$0	$0
Pipeline Recoating	($2,000,000)	$0	$0	$0	$0	$0	$0	$0	$0	$0	$0	$0	$0	$0	$0	$0
Incremental Cash Flow	*($1,550,000)*	*$500,000*	*$500,000*	*$500,000*	*$500,000*	*$500,000*	*$500,000*	*$500,000*	*$500,000*	*$500,000*	*$500,000*	*$0*	*$0*	*$0*	*$0*	*$0*

ROI measures for Option One	
Cost of capital	11%
Net present value (NPV)	*$1,256,411*

Option Two: Replacement	Year 0	Year 1	Year 2	Year 3	Year 4	Year 5	Year 6	Year 7	Year 8	Year 9	Year 10	Year 11	Year 12	Year 13	Year 14	Year 15
Asset Cash Flow Generation	$500,000	$500,000	$500,000	$600,000	$600,000	$600,000	$600,000	$600,000	$600,000	$600,000	$600,000	$600,000	$600,000	$600,000	$600,000	$600,000
Pipeline Replacement	$0	$0	$0	(5,000,000)	$0	$0	$0	$0	$0	$0	$0	$0	$0	$0	$0	$0
Incremental Cash Flow	*$500,000*	*$500,000*	*$500,000*	*($4,400,000)*	*$600,000*	*$600,000*	*$600,000*	*$600,000*	*$600,000*	*$600,000*	*$600,000*	*$600,000*	*$600,000*	*$600,000*	*$600,000*	*$600,000*

ROI measures for Option Two	
Cost of capital	11%
Net present value (NPV)	*$889,471*

APPENDIX B

Economics of Corrosion

Much of the information in Appendix B has been extracted from NACE International Publication 3C194.[16]

More than $276 billion are lost to corrosion in the United States each year, according to government and industry studies.[4] This corrosion cost is equivalent to at least three percent of the gross domestic product (GDP). The real tragedy of this annual corrosion cost is that it has been estimated that 25% to 30% of annual corrosion costs in the U.S. could be saved by the application of existing technology to prevent and control corrosion.[4] The existing technology includes the following methodologies to prevent and control corrosion: proper design, selection of materials, coatings and linings, CP, and inhibitors. Standards, reports, books, and thousands of technical articles attest to the successful use of the existing technology to prevent and control corrosion and thus to reduce the annual losses to corrosion.

Corrosion—An Economic Problem

Corrosion is essentially an economic problem. It is vitally important that engineers and engineering managers be aware of the economic impact their decisions have on the ability of a business to meet its corporate goals. A discipline that assists in the measurement of the economic impact of such decisions is called *engineering economy*.

Engineering economy is concerned with money, both as a resource and as the price of other resources. Business success is dependent on the prudent and efficient use of all resources, including money. The principles of engineering economy permit direct comparisons of potential alternatives in monetary terms. In this way they encourage efficient use of resources.

The economics of corrosion evaluation involves the assessment of the technical validity and economic justification of each alternative. Corrosion technologists generally spend much effort exploring the technical validity of materials or processes. However, each corrosion problem may have more than one material or process that could satisfactorily solve the technical problem. Each candidate material or process probably has a unique stream of investment and operating or maintenance costs, while providing equivalent technical benefits. Indeed, once the technically viable alternatives are chosen, and their respective costs and performance characteristics are isolated, the technologist's decision becomes one of financial analysis, because almost all of the factors in the decision process can be reduced to the magnitude and timing of cash flows.

The purposes of this discussion are to

1. Present the economic techniques in a form that can be readily understood and used by engineers as a decision-making tool.
2. Facilitate the communication of decisions between the corrosion technologist and management.
3. Justify investment in anti-corrosion methods that have long-term benefits.

Basics—Money and Time

Consider the effects of time and earning power on \$20.00. If \$20.00 is placed in the bank and earns interest at a rate of 5% per year, it grows to \$21.00 when the interest, \$1.00, is paid at the end of the year. Thus, \$20.00 today at 5% is equivalent to \$21.00 a year from now. Stated another way, in order to have \$21.00 one year from now, only \$20.00 has to be deposited today if the interest rate is 5%. The \$20.00 is called the "discounted present value" of the \$21.00 needed one year hence.

The initial deposit as well as the earned interest left in the account have earning power because the interest is compounded, which means that it is computed on both the principal and the accrued interest.

Suppose the \$20.00 were used to pay four equal annual installments of \$5.00 each. Without interest, the \$20.00 would be exhausted after making the last payment. If the \$20.00 is deposited in the bank at 5% interest, it will be worth \$21.00 at the end of the first year. Paying out \$5.00 would leave \$16.00 to be held at interest the second year. \$16.00 invested at 5% will earn \$0.80 in one year. Subtracting \$5.00 from \$16.80 leaves \$11.80 to be held at 5% interest for the third year. \$11.80 will earn \$0.59 by the end of the third year at 5%. Another annual payment of \$5.00 leaves \$7.39 to earn interest during the fourth year. Adding the $0.37 interest earned during the fourth year and subtracting the final $5.00 annual payment leaves a balance of accrued interest of $2.76.

The earning power of money permits another strategy. If the initial deposit were reduced to $17.73 at 5% interest, $5.00 could be paid out each year for four years and nothing would be left. This example illustrates the distinction between the terms "equivalent" and "equal." $20.00 is equal to four payments of $5.00 each. It also would be equivalent to four $5.00 annual payments (only) if the interest rate were zero. The $17.73 is not equal to the sum of four payments of $5.00 each. However, when $17.73 is invested at 5%, it is equivalent to four annual payments of $5.00.

The term "equivalent" implies that the concept of the time-value of money is applied at some specific interest rate. Therefore, for an amount of money to have a precise meaning, it must be fixed both in time and amount. Mathematical formulae and tables are available to translate an amount of money at any particular time into an equivalent amount at another date.

Many kinds of translations are possible. For example, a single amount of money can be translated into an equivalent amount at either a later or an earlier date. This is accomplished by calculating the "present worth" (PW) or the "future worth" (FW) as of the present date. Single amounts of money can be translated into equivalent "annuities" (A) involving a series of uniform amounts occurring each year. Conversely, annuities can be translated into equivalent single amounts at either an earlier or a later date. The "present worth of an annuity" (P/A) is the single amount of money equivalent to a future annuity. The single amount equivalent to a past annuity is referred to as the "future worth of an annuity" (F/A).

It is also possible to calculate the amount of money that would be equivalent to a nonuniform series of cash flows. Two types of nonlinear series that find application are an arithmetic progression, in which the series changes by a constant amount, and a geometric progression, in which the series changes by a constant rate. The arithmetic progression is considered to be representative of variable costs such as maintenance costs, which may increase as equipment ages. The geometric progression is used to represent the effects of inflation or deflation.

Terminology

The definitions of technical terms used in engineering economy are available on the Institute of Industrial Engineers (IIE) Web site at: http://www.iienet2.org/Details.aspx?id=1902.

Depreciation

Depreciation has been defined as the lessening in value of an asset with the passage of time. All physical assets (with the possible exception of land) depreciate with time. There are several types of depreciation. Two of the more common types are physical depreciation and functional depreciation. Accidents can also cause loss of value, but this cause is often accommodated in other ways (i.e., insurance, reserves, etc.) and is not considered here.

Physical depreciation includes such phenomena as deterioration resulting from corrosion, rotting of wood, bacterial action, chemical decomposition, wear and tear, etc., which can reduce the ability of an asset to render its intended service.

Functional depreciation results not from the inability of an asset to serve its intended purpose, but rather from the fact that some other asset that can perform the desired function more economically is available. Thus obsolescence and/or inadequacy, or inability to meet the demands placed on the asset, lead to functional depreciation. Technological advances produce improvements that often result in obsolescence of existing assets.

The manner in which depreciation is accounted for is largely a tax question. The language of the tax laws specifies the procedures that are permissible. Tax laws change from time to time, and so procedures that are attractive under a given set of circumstances may become unattractive (or even forbidden) under other circumstances.

Worked Examples and Applications

Problem 1—Process Equipment

A new heat exchanger is required in conjunction with rearrangement of existing facilities. Because of corrosion, the expected life of a carbon steel exchanger is five years. The installed cost is $9,500. It is proposed to substitute a UNS S31600 (AISI 316 SS) exchanger having an installed cost of $26,500, and an estimated life of 15 years, written off in 11 years. Which is the more economical choice based on annual cost? The minimum acceptable rate of return is 10%, the tax rate is 48%, and the depreciation method is straight-line.

Because the lives of the alternatives are unequal, the economic choice cannot be based merely on the discounted cash flow over a single life of each alternative. Instead, comparison is made on the basis of equivalent uniform annual costs, commonly referred to as "annual cost," as mentioned above. Referring to the generalized equation shown above, in this problem, data are available for only the first two terms of the equation. The third term, which involves maintenance expense, and the fourth term, involving salvage value, are both assumed to be zero.

The first step is to compute the discounted cash flow over one life span for each alternative. This involves the calculation of the present worth (PW) for each material (Equations [B1] and [B2]).

$$(PW)_{steel} = -\$9{,}500 + \left\{\frac{0.48\,(9{,}500 - 0)}{5}\right\}(3.791) = -\$6{,}043 \quad \text{(B1)}$$

$$(PW)_{316} = -\$26{,}500 + \left\{\frac{0.48\,(26{,}500 - 0)}{11}\right\} \quad \text{(B2)}$$

To compare the two alternatives, the discounted cash flows (PWs) each are converted to annual costs (Equations [B3] and [B4]).

$$A_{steel} = -\$6{,}043\,(0.2638) = -\$1{,}594 \quad \text{(B3)}$$

$$A_{316} = -\$18{,}989\,(0.1540) = -\$2{,}924 \quad \text{(B4)}$$

Therefore, the carbon steel heat exchanger, which has the lower annual cost, is the more economical alternative under these conditions.

Problem 1a—If the carbon steel exchanger were to require \$3,000 yearly maintenance (e.g., painting, use of inhibitors, cathodic protection, etc.), would carbon steel still be the preferred alternative?

Maintenance costs are treated as expense items with one-year life (end-of-year costs). This particular exercise involves the third term in the generalized equation shown above. In Problem 1, the annual cost without the yearly maintenance cost has already been computed, so in this case the annual cost including maintenance (stated to be \$3,000 per year) is obtained simply by subtracting the after-tax maintenance costs from the result given in Problem 1 (Equation [B5]).

$$A_{steel} = -\$1{,}594 - (1 - 0.48)(\$3{,}000) = -\$3{,}154 \quad \text{(B5)}$$

Because this is an after-tax cash flow, the quantity [(1 − 0.48) (\$3,000)] is the after-tax maintenance cost where 0.48 equals the tax rate expressed as a decimal. The \$3,000 is the present value of the first year's maintenance, so no further discounting is necessary in this case.

Comparing this result with the annual cost of UNS S31600 in Problem 1, it is evident that if \$3,000 is required each year to keep the carbon steel unit operative, the UNS S31600 unit would be more economical.

Problem 1b—Under the conditions described in Problem 1, if it is not certain that a five-year life will be attained for unprotected carbon steel, at what life (for carbon steel) is it economically equivalent to the UNS S31600 heat exchanger?

The carbon steel heat exchanger will be economically equivalent to the UNS S31600 heat exchanger when their annual costs are equal. Therefore, the problem can be solved by determining how many years' life for steel is equivalent to an annual cost of $2,924. Trial-and-error methods are useful for such problems.

Trying n = 3 years (Equations [B6] and [B7]):

$$(PW)_{steel} = -\$9{,}500 + 0.48\left\{\frac{(9{,}500 - 0)}{3}\right\}(2{,}487) = -\$5{,}719 \quad \text{(B6)}$$

$$A_{steel} = -\$5{,}719\,(0.40211) = -\$2{,}300 \quad \text{(B7)}$$

Trying n = 2 years (Equations [B8] and [B9]):

$$(PW)_{steel} = -\$9{,}500 + 0.48\left\{\frac{(9{,}500 - 0)}{2}\right\}(1{,}736) = -\$5{,}542 \quad \text{(B8)}$$

$$A_{steel} = -\$5{,}542\,(0.57619) = -\$3{,}319 \quad \text{(B9)}$$

Thus a carbon steel heat exchanger must last more than two years but is economically favored in less than three years under the conditions given.

Problem 1c—Under the described conditions, how much product loss, "X," could be tolerated after two of the five years of anticipated life (e.g., from roll leaks or a few tube failures) before the selection of UNS S31600 is justified?

Equating (Equations [B10] and [B11]):

$$A_{316} = A_{steel} + A_{product\ loss} \quad \text{(B10)}$$

$$-2{,}924 = -1{,}594 + [(1 - 0.48)]\,(X)\,[(0.8264)]\,[0.2638] \quad \text{(B11)}$$

Solving for X (Equations [B12] and [B13]):

$$-1{,}430 = 0.1134\,(X) \quad \text{(B12)}$$

$$X = -\$12{,}610 \quad \text{(B13)}$$

The term (1 – 0.48) is recognized from term three of the general equation as $(1 - t)$ wherein "t" is the tax rate. The quantity 0.8264 is the single-payment present worth factor (P/F, 10%, two years) and translates the product loss, "X," to its present worth at time zero. The quantity [0.2638] is the uniform series capital recovery factor (A/P, 10%, five years) and translates the present worth of the product loss to annual cost so that it can be added to the annual cost of steel for comparison with the annual cost of UNS S31600. If production losses exceed $12,610 in year two (no losses in other years), the UNS S31600 heat exchanger is justified.

Problem 2—Cathodic Protection

It is proposed to cathodically protect an underground piping installation, assuming the following: i = 10%, t = 34%, straight-line depreciation. Three proposals have been made:

1. Anodes, 4,250 at $100,000, installed. Useful life, 10 years.
2. Thirty (30) rectifiers with groundbeds at $100,000, installed. Expected life, 20 years with annual maintenance cost of $5,900.
3. A "mixed" system involving rectifiers ($18,000 installed plus $1,200/y maintenance for 20-year life) plus anodes ($82,000 installed for 10-year life).

For Proposal 1 (see Equations [B14] and [B15]):

$$(PW)_1 = -100{,}000 + (0.34)\left\{\frac{(100{,}000 - 0)}{10}\right\}(6.1446) = -\$79{,}108 \tag{B14}$$

$$A_1 = -79{,}108\,(0.1628) = -\$12{,}879 \tag{B15}$$

For Proposal 2 (see Equations [B16] and [B17]):

$$(PW)_2 = -100{,}000 + (34)\left\{\frac{(100{,}000 - 0)}{20}\right\}(8.5136)$$
$$- (1 - 0.34)\,(5{,}900)\,(8.5136) = -118{,}679 \tag{B16}$$

$$A_2 = -118{,}679\,(0.1175) = -\$13{,}945 \tag{B17}$$

For Proposal 3.

For the anode portion, see Equations (B18) and (B19):

$$(PW)_3 = -82{,}000 + (0.34)\frac{(82{,}000 - 0)}{10}(6.1446) = -\$64{,}869 \tag{B18}$$

$$A_3 = -118{,}679\,(0.1628) = -\$10{,}561 \tag{B19}$$

For the rectifier-driven portion, see Equations (B20), (B21), and (B22):

$$(PW)_3 = -18{,}000 + (0.34)\frac{(18{,}000 - 0)}{20}(8.5136) - (1 - 34)(1{,}200)(8.5136) = -22{,}138 \tag{B20}$$

$$A_3 = -22{,}138\,(0.1175) = -\$2{,}601 \tag{B21}$$

$$A_T = A_3\ anodes + A_3\ rectifiers = -10{,}561 - 2{,}601 = -\$13{,}162 \tag{B22}$$

Under these particular conditions, Proposal 1 is calculated to be the least expensive alternative on the basis of lowest annual cost, followed by Proposal 3.

APPENDIX C

Select Aboveground Coating Standards

TABLE C.1 Common Coatings for Submerged Pipelines

Coating Type	Relevant Standards	Description/Characteristics
Fusion-Bonded Epoxy (FBE)	NACE Standard RP0394, Application, Performance, and Quality Control of Plant-Applied, Fusion-Bonded Epoxy External Pipe Coating[69] NACE Standard RP0402, Field-Applied Fusion-Bonded Epoxy (FBE) Pipe Coating Systems for Girth Weld Joints: Application, Performance, and Quality Control[70] API RP 5L7, Unprimed Internal Fusion Bonded Epoxy Coating of Line Pipe[71] CSA[(10)] Z245.20-M92, External Fusion Bonded Epoxy Coating for Steel Pipe[72] ANSI/AWWA[(11)] C 213, Fusion-Bonded Epoxy Coating for the Interior and Exterior of Steel Water Pipelines[73] ISO 21809-2, External coatings for buried or submerged pipelines used in pipeline transportation systems—Part 2: Fusion-bonded epoxy coatings[74]	FBE is a high-integrity coating with good bonding strength, good dielectric properties, and displays good durability and resistance to degradation. A detailed and thorough application process involving pre-heating, cleaning, grinding, and application and curing is involved.[75] These coatings are typically polymer coatings that have a chemical cross-linking. May be applied in single, double, and triple layers in accordance with the intended service environment.

(continued)

[(10)] CSA International (CSA), 178 Rexdale Blvd., Toronto, Ontario M9W, 1R3, Canada.
[(11)] American Water Works Association (AWWA), 6666 W. Quincy Ave., Denver, CO 80235.

TABLE C.1 *(continued)*

Coating Type	Relevant Standards	Description/Characteristics
Liquid Epoxy and Polyurethane	NACE Standard RP0105, Liquid-Epoxy Coatings for External Repair, Rehabilitation, and Weld Joints on Buried Steel Pipelines[76] ISO 21809-3, External coatings for buried or submerged pipelines used in pipeline transportation systems—Part 3: Field joint coatings[77]	Also considered a high-integrity coating with good bonding strength and resistance to deterioration. Applied in a liquid form, this coating covers up the exterior of the structure as one continuous layer. The application of these types of coatings creates a strong, thin, water-resistant film over the surface of the pipeline with various film thickness possibilities.
Extruded Polyolefin	NACE Standard SP0185, Extruded Polyolefin Resin Coating Systems with Soft Adhesives for Underground or Submerged Pipe[78] ISO 21809-4, External coatings for buried or submerged pipelines used in pipeline transportation systems—Part 4: Polyethylene coatings (2-layer PE)[79]	An extruder heats, melts, mixes, and extrudes the material onto the steel pipe. A copolymer adhesive is typically applied followed by application of an epoxy primer, such as asphaltic mastic.
Multi-Layer Coatings	ISO 21809-3, External coatings for buried or submerged pipelines used in pipeline transportation systems—Part 1: Polyolefin coatings (3-layer PE and 3-layer PP)[80]	Coating typically consists of an inner layer (such as epoxy or FBE) and outer layer (such as polyolefin) held together by an adhesive.
Coal Tar Coatings	NACE Standard RP0399, Plant-Applied External Coal Tar Enamel Pipe Coating Systems: Application, Performance, and Quality Control[81] NACE Standard RP0602, Field-Applied Coal Tar Enamel Pipe Coating Systems: Application, Performance, and Quality Control[82]	Coal tar is a material derived from coking and processing of coal, which is then blended with inert fillers or other materials to form the coal tar coating. The coal tar is usually applied as a hot melt to the exterior of the pipe, usually followed by a fibrous wrap to ensure smooth application and remove air pockets.

Coating Type	Relevant Standards	Description/Characteristics
	ANSI/AWWA C 203, AWWA Standard for Coal-Tar Protective Coatings and Linings for Steel Water Pipelines Enamel and Tape Hot Applied[83] SSPC-Paint 32, Coal Tar Emulsion Coating[84] SSPC-Paint 33, Coal Tar Mastic, Cold Applied[85] SSPC Paint 16, Coal Tar Epoxy Black (or Dark Red) Paint[86]	
Tape or Wrap Coatings (including Polyethylene)	CSA Z245.21-M92, External Polyethylene Coating for Pipe[87] ANSI/AWWA C214, Standard for Tape-Coating Systems for the Exterior of Steel Water Pipelines[88] ISO 21809-3, External coatings for buried or submerged pipelines used in pipeline transportation systems—Part 3: Field joint coatings	Polyethylene and other tapes have been wrapped over pipelines with the intention of separating the steel from the electrolyte. The system typically consists of a primer, a corrosion-preventive layer, and one or two outer layers for mechanical protection.
Shrink-Sleeve	NACE Standard RP0303, Field-Applied Heat-Shrinkable Sleeves for Pipelines: Application, Performance, and Quality Control[89] ISO 21809-3, External coatings for buried or submerged pipelines used in pipeline transportation systems—Part 3: Field joint coatings	Typically a thermoset plastic that can be applied to the pipeline and when heated will shrink around the pipe. Often used for pipeline joints. They are prepackaged in a blanket form that is wrapped around the pipe and shrinks to it when heated.

(*continued*)

TABLE C.1 (*continued*)

Coating Type	Relevant Standards	Description/Characteristics
Wax Coatings	NACE Standard RP0375, Field-Applied Underground Wax Coating Systems for Underground Pipelines: Application, Performance, and Quality Control[90] ISO 21809-3, External coatings for buried or submerged pipelines used in pipeline transportation systems—Part 3: Field joint coatings[87]	Wax coatings include hot-applied waxes, cold-applied waxes, component wrappers, and wax tapes. Hot-applied waxes are typically composed of microcrystalline wax and sometimes contain wetting agents, fillers, and corrosion inhibitors. Waxes are often applied to irregularly shaped surfaces.
Self-Healing Visco-Elastic Coating Systems	ISO 8504-3, Preparation of steel substrates before application of paints and related products—Surface preparation methods—Part 3: Hand- and power-tool cleaning[91]	The system consists of a non-crystalline (fully amorphous, non-crosslinked) polyolefin-based corrosion-preventive layer with an outer layer for mechanical protection (such as polymeric tapes, heat-shrink sleeves, glass fiber-reinforced epoxy, polyurethane, flamespray, polyester or vinyl ester wrap) suited to the particular application environment. This type of coating system, which has a glass transition temperature of less than −60 °C, can be applied cold with minimal surface preparation (ISO ST 2/3) and exhibits immediate molecular adhesion to any substrate (e.g., polyethylene, polypropylene, FBE). It is 100% self-healing after hot-water immersion testing for 100 days at 90 °C. There is no need for blasting, it is applied without a primer, and it has no cathodic disbondment, no CP shielding, and no MIC or osmosis.

TABLE C.2 Coating Type, Standards, and Considerations

Coating Type	SSPC and Other Standards	Characteristics/Notes
Baked Phenolic	NA	• Excellent resistance to acidic environments/water • Low material costs/high application costs • Not normally used on external pipeline surfaces
Epoxy	SSPC-Paint 34, W[92] ANSI/AWWA C210, Standard for Liquid-Epoxy Coating Systems for the Interior and Exterior of Steel Water Pipelines[93]	• Three generic types: oil modified, catalyzed, and high-baked • Varying chemical resistance based on type • Typically strong vapor barrier
Vinyl	SSPC-PS 4.04, Four-Coat White or Colored Vinyl Painting System[94] SSPC-PS 4.02, Four-Coat Vinyl Painting System[95] SSPC-PS Guide 4.00, Guide for Selecting Vinyl Painting Systems[96] SSPC-Paint 27, Basic Zinc Chromate-Vinyl Butyral Wash Primer[97] SSPC-Paint 9, White (or Colored) Vinyl Paint[98] SSPC-Paint 8, Aluminum Vinyl Paint[99] SSPC-Paint 106, Black Vinyl Paint[100]	• Greatest versatility of commonly used resin types • Good pH and water resistance • May have limited solvent and heat resistance • Does not meet North American and European volatile emissions (VOC) regulations
Zinc (Inorganic and Organic)	SSPC-Paint 30, Weld-Through Inorganic Zinc Primer[101] SSPC-Paint 20, Zinc-Rich Coating (Type 1—Inorganic, and Type II—Organic[102] SSPC-Guide 14, Guide for the Repair of Imperfections in Galvanized, Organic, or Inorganic Zinc-Coat[103] SSPC-Paint 29, Zinc Dust Sacrificial Primer, Performance-Based[104]	

(*continued*)

TABLE C.2 *(continued)*

Coating Type	SSPC and Other Standards	Characteristics/Notes
	SSPC PS Guide 12.00, Guide to Zinc-Rich Coating Systems[105] SSPC-PS Guide 8.00, Guide to Topcoating Zinc-Rich Primers[106] SSPC-PS 15.04, Chlorinated Rubber Painting System for Field Application over a Shop-Applied Solvent[107] SSPC-Paint 30, Weld-Through Inorganic Zinc SSPC-PS Guide 8.00, Guide to Topcoating Zinc-Rich Primers SSPC-Guide 20.00, Guide for Selecting Painting Systems for Boottoppings[108]	• Some of the strongest weather and solvent resistance • Successfully used to protect against corrosion in all types of exposed steel • Metallic zinc dissolves anodically to steel, providing a form of CP
Furan	NA	• Limited acceptance because of application difficulties • Very versatile and resistant to aggressive chemicals • Difficult to maintain • Rarely used on external pipeline surfaces
Coal Tar Epoxy	SSPC-Paint 32, Coal-Tar Emulsion Coating SSPC-Paint 33, Coal-Tar Mastic Coating, Cold-Applied NACE Standard RP0602, Field-Applied Coal Tar Enamel Pipe Coating Systems: Application, Performance, and Quality Control SSPC-Paint 16, Coal Tar Epoxy Black (or Dark Red) Paint	• Mix of coal-tar pitch and low-molecular-weight epoxy resins • Very resistant to acids and water • High chalking rate when exposed to sunlight (more frequently used on buried pipeline)
Thermal Spray (Metallizing)	NACE No. 12/SSPC-CS23.00/AWS C2.23M/, Specification for the Application of Thermal Spray Coatings (Metallizing) of Aluminum, Zinc, and Their Alloys and Composites for the Corrosion Protection of Steel[109]	• Thermal sprayed zinc or zinc/aluminum mixture (TSZ) preferred for ambient services. Thermal spray aluminum (TSA) has superior corrosion resistance in immersion. • Normally sealed with thin-film epoxy, vinyl, or silicone in immersion service

Coating Type	SSPC and Other Standards	Characteristics/Notes
Polyurethane/ Polyurea (Elastomeric and Rigid Film)	ANSI/AWWA C222, Standard for Polyurethane Coatings for the Interior and Exterior of Steel Water Pipe and Fittings[110]	• Nearly unlimited film thickness in one application • Sprayable down to −200 °F (−35 °C) • Excellent abrasion resistance • Wide range of formulations for specific service environments • Applicable over other generic coatings for girth weld repairs • Requires plural component heated airless equipment for optimum performance
Polysiloxane	Norsok M501, Surface preparation and protective coating[111] ASTM D5894, Standard Practice for Cyclic Salt Fog/UV Exposure of Painted Metal (alternating Exposures in a Fog/Dry Cabinet and a UV/Condensation Cabinet[112] ISO 12944, Paints and varnishes—Corrosion protection of steel structures by protective paint systems[113]	• Epoxy and polyurethane resin backbone types available • May be applied direct to metal or over inorganic and organic primers. • Normally limited to thin films in ambient service • Excellent color and gloss retention in ambient service • Flexibility is formula dependent

APPENDIX D

Terms and Definitions

Assessment (Pipeline Assessment): Comprehensive process or processes incorporating various inspection tools and techniques to collect, analyze, and disposition data to determine the state or condition of a pipeline segment.

Cathodic Protection: A technique to reduce the corrosion of a metal surface by making that surface the cathode of an electrochemical cell.

Cleaning Pig: A device inserted in a pipeline for cleaning solids or displacing liquids or hydrocarbons from within a pipeline.

Coating Holiday or Holiday: A discontinuity in a protective coating that exposes unprotected surface to the environment.

Corrosion: The deterioration of a material, usually a metal, that results from a chemical or electrochemical reaction with its environment.

Corrosion Risk: Potential impact composed of the likelihood and consequence of damage, injury, liability, loss, or other negative occurrences caused by a pipeline failure from corrosion.

Cost Benefit Analysis: An analysis of the cost effectiveness of different alternatives to see whether the benefits outweigh the costs.[114]

Direct Assessment: A structured process for pipeline operators to assess the integrity of pipelines.

External Corrosion: Corrosion occurring on the outside wall of a pipe.

External Corrosion Direct Assessment: A four-step process that combines preassessment, indirect inspection, direct examination, and postassessment to evaluate the effect of corrosion occurring on the outside wall of a pipe on the integrity of a pipeline.

Environmentally Assisted or Induced Cracking: Cracking of a material wherein an interaction with its environment is a causative factor in conjunction with tensile stress, often resulting in brittle fracture of an otherwise ductile material.

Hydrostatic Pressure Testing (Hydrotest): Testing of sections of a pipeline by filling the pipeline with water and pressurizing it until the nominal hoop stresses in the pipe reach a specified value.

In-Line Inspection (ILI or Smart Pigging): The inspection of a pipeline from the interior of the pipe using an in-line inspection (ILI) tool. The tools used to conduct ILI are known as pigs, smart pigs, or intelligent pigs.

Internal Corrosion (IC): Corrosion occurring on the interior wall of a pipeline.

Internal Corrosion Direct Assessment: A four-step process that combines preassessment, indirect inspections, direct examinations, and postassessment to evaluate the impact of corrosion occurring on the inside wall of a pipe on the integrity of a pipeline.

Dry Gas Internal Corrosion Direct Assessment: The internal corrosion direct assessment process as defined in NACE SP0206 applicable to normally dry gas systems.

Hot Tapping (Pressure Tapping): The method of making a connection or tie into existing piping or vessels without interruption to the system.

Liquid Petroleum Internal Corrosion Direct Assessment: The internal corrosion direct assessment process as defined in NACE SP0208 applicable to liquid petroleum systems.

Mitigation: The effort to reduce the risk of a pipeline failure from corrosion through a defined set of actions.

Net Present Value: Difference between the present value (PV) of the future cash flows from an investment and the amount of investment. Present value of the expected cash flows is computed by discounting them at the required rate of return (also called minimum rate of return).

PH: The negative logarithm of the hydrogen ion activity written as

$$pH = -\log_{10}({}^{a}{}_{H^+})$$

where

${}^{a}{}_{H^+}$ = hydrogen ion activity = the molar concentration of hydrogen ions multiplied by the mean ion-activity coefficient.

Pipeline Corrosion Management: A comprehensive program for reducing the effects of corrosion on a transmission pipeline to an acceptable level.

Pipeline Threat: One of 22 hazard classifications or 9 prescriptive threats that adversely affect the safe and reliable operation of a pipeline. For the purposes of this guide, only the three time-dependent prescriptive corrosion threats are considered in assessing the integrity of the pipeline.

Potential Pipeline Threat: A potential hazard that adversely affects the safe and reliable operation of a pipeline.

Remediation: Corrective actions taken to mitigate or reduce failure likelihood from a corrosion threat.

Stress: The force per unit area when a force acts on a body.

Stress Corrosion Cracking: Cracking of a material produced by the combined action of corrosion and sustained tensile stress (residual or applied).

Stress Corrosion Cracking Direct Assessment: A four-step integrity methodology that combines preassessment, indirect inspections, direct examinations, and postassessment to evaluate the possibility of SCC and estimate the impact of deterioration on the outside wall of a pipe to reduce the integrity of a pipeline and estimate the interval between integrity assessments.

Tensile Stress: Stress that tends to elongate the material.

Yield Strength: The stress at which a material exhibits a specified deviation from the proportionality of stress to strain. The deviation is expressed in terms of strain by either the offset method (usually at a strain of 0.2%) or the total-extension-under-load-method (usually at a strain of 0.5%).

Index